Frank Heberling

Structural Incorporation of Neptunyl(V) into Calcite

Frank Heberling

Structural Incorporation of Neptunyl(V) into Calcite

Interfacial Reactions and Kinetics

Südwestdeutscher Verlag für Hochschulschriften

Impressum/Imprint (nur für Deutschland/ only for Germany)
Bibliografische Information der Deutschen Nationalbibliothek: Die Deutsche Nationalbibliothek verzeichnet diese Publikation in der Deutschen Nationalbibliografie; detaillierte bibliografische Daten sind im Internet über http://dnb.d-nb.de abrufbar.

Alle in diesem Buch genannten Marken und Produktnamen unterliegen warenzeichen-, marken- oder patentrechtlichem Schutz bzw. sind Warenzeichen oder eingetragene Warenzeichen der jeweiligen Inhaber. Die Wiedergabe von Marken, Produktnamen, Gebrauchsnamen, Handelsnamen, Warenbezeichnungen u.s.w. in diesem Werk berechtigt auch ohne besondere Kennzeichnung nicht zu der Annahme, dass solche Namen im Sinne der Warenzeichen- und Markenschutzgesetzgebung als frei zu betrachten wären und daher von jedermann benutzt werden dürften.

Verlag: Südwestdeutscher Verlag für Hochschulschriften Aktiengesellschaft & Co. KG
Dudweiler Landstr. 99, 66123 Saarbrücken, Deutschland
Telefon +49 681 37 20 271-1, Telefax +49 681 37 20 271-0
Email: info@svh-verlag.de
Zugl.: Karlsruhe, KIT, Diss., 2010

Herstellung in Deutschland:
Schaltungsdienst Lange o.H.G., Berlin
Books on Demand GmbH, Norderstedt
Reha GmbH, Saarbrücken
Amazon Distribution GmbH, Leipzig
ISBN: 978-3-8381-1527-6

Imprint (only for USA, GB)
Bibliographic information published by the Deutsche Nationalbibliothek: The Deutsche Nationalbibliothek lists this publication in the Deutsche Nationalbibliografie; detailed bibliographic data are available in the Internet at http://dnb.d-nb.de.

Any brand names and product names mentioned in this book are subject to trademark, brand or patent protection and are trademarks or registered trademarks of their respective holders. The use of brand names, product names, common names, trade names, product descriptions etc. even without a particular marking in this works is in no way to be construed to mean that such names may be regarded as unrestricted in respect of trademark and brand protection legislation and could thus be used by anyone.

Publisher: Südwestdeutscher Verlag für Hochschulschriften Aktiengesellschaft & Co. KG
Dudweiler Landstr. 99, 66123 Saarbrücken, Germany
Phone +49 681 37 20 271-1, Fax +49 681 37 20 271-0
Email: info@svh-verlag.de

Printed in the U.S.A.
Printed in the U.K. by (see last page)
ISBN: 978-3-8381-1527-6

Copyright © 2010 by the author and Südwestdeutscher Verlag für Hochschulschriften Aktiengesellschaft & Co. KG and licensors
All rights reserved. Saarbrücken 2010

Abstract

The disposal of high–level nuclear waste in deep geological formations poses major scientific and social challenges to be met in the next decades. One of the key issues is the long term safety of a waste repository over extended periods of time, up to 10^6 years. Over geological time spans it cannot be excluded that groundwater will infiltrate into a waste repository. It needs to be considered that permeating groundwater corrodes waste canisters and dissolves and thereby mobilizes radionuclides from the waste. Demonstrating the repository safety over geological time spans requires a sound understanding of the migration and retention behavior of radionuclides in the geosphere. In principle, sorption reactions between radionuclides and minerals control the retention. In recent years, various molecular-level sorption mechanisms have been identified: e.g. outer-sphere and inner-sphere adsorption, ion exchange, and structural incorporation (solid solution formation).

The actinide elements are important radionuclides in studies related to high level nuclear waste disposal due to their major contribution to the long term radiotoxicity of spent nuclear fuel. In oxidizing aqueous environments the actinide elements uranium, neptunium, and plutonium form linear trans–dioxo cations, the so called actinyl ions: $U(VI)O_2^{2+}$, $Np(V)O_2^+$, and $Pu(V,VI)O_2^{+/2+}$. In this experimental study sorption reactions of neptunyl(V), $Np(V)O_2^+$, with the mineral phase calcite have been investigated. The calcite surface is investigated in contact to solution of various compositions by means of zetapotential and surface diffraction measurements. Calcium and carbonate ions are determined as the the potential determining ions at the calcite surface while pH has only a minor influence on the zetapotential. No indication for inner–sphere complexes at basal planes on the calcite surface has been found. Instead two well ordered layers of water have been identified at 2.35 ± 0.05 Å and 3.24 ± 0.06 Å above the surface. A Basic–Stern surface complexation model is presented that that describes the measured zetapotentials considering only outer–sphere complexes of ions other than protons and hydroxide.

The adsorption of neptunyl(V) at calcite is investigated over a large range of neptunyl concentrations and CO_2 partial pressures. Adsorption shows a "Freundlich"–like concentration dependence and is pH dependent up to an equilibrium concentration of 10^{-5} M NpO_2^+, with maximum adsorption at pH 8.3. The highest surface loading observed, corresponds to 2 % surface site coverage. At pH 8.3 and high surface coverages a bidentate inner–sphere adsorbed biscarbonato complex bound to step edge sites on the calcite (104)–face is indentified by low temperature (15K) EXAFS as the most abundant adsorption species.

Upon coprecipitation neptunyl(V) is readily incorporated into the calcite structure. EXAFS investigations identify an incorporation species with a equatorial coordination environment of four monodentate bound carbonate ions around the central linear neptunyl moiety. Neptunyl likely substitutes one calcium and two adjacent carbonate ions in the calcite host. How the charge excess introduced by this substitution is balanced is not known. The retardation of calcite crystal growth due to neptunyl coprecipitation supports the idea of adsorption complexes at step edge sites. Experiments on adsorption kinetics and on desorption indicate that even at calcite equilibrium conditions incorporation reactions might take place, to a limited extent.

The reactivity of the calcite surface, the high affinity of neptunyl(V) for adsorption onto the calcite surface, and the propensity of neptunyl(V) to be incorporated into the calcite structure, make calcite a potentially important sink for neptunyl(V) in the geosphere.

Zusammenfassung

Die Endlagerung hoch–radioaktiver nuklearer Abfälle in tiefen geologischen Formationen stellt eine wissenschaftliche und gesellschaftliche Herausforderung dar, die es in den nächsten Jahrzehnten zu bewältigen gilt. Unsicherheiten sind hauptsächlich mit der Langzeitsicherheit eines Endlagers über ausgedehnte Zeiträume von bis zu 10^6 Jahren verbunden. Es kann nicht ausgeschlossen werden, dass über geologische Zeiträume Grundwasser in ein Endlager eindringt. Dabei muss in Betracht gezogen werden, dass eintretendes Grundwasser Abfallbehälter korrodiert, Radionuklide aus der Abfallmatrix löst und dadurch mobilisiert. Ein fundiertes Verständnis der Transport– und Rückhaltemechanismen für Radionuklide in der Geosphäre ist Voraussetzung dafür, die mit der Endlagerung verbundenen Risiken über geologische Zeiträume abschätzen zu können. Sorptionsreaktionen mit den umgebenden Minaralphasen sind ausschlaggebend für die Rückhaltung von Radionukliden in der Geosphäre. Einige Sorptionsmechanismen sind in den vergangenen Jahren im molekularen Maßstab charakterisiert worden: z.B. außersphärische und innersphärische Adsorption, Ionenaustausch und struktureller Einbau (Bildung von Mischkristallen). In Untersuchungen zur Langzeitsicherheit von Endlagern für hochradioaktive Nukleare Abfälle spielen die Actiniden mit die wichtigste Rolle, da sie über lange Zeiträume einen Großteil des Radiotoxizitätsinventars abgebrannten Kernbrennstoffes ausmachen. In oxidierenden wässrigen Lösungen liegen die Actiniden–Elemente Uran, Neptunium und Plutonium als Trans–Dioxo–Kationen, $U(VI)O_2^{2+}$, $Np(V)O_2^+$, and $Pu(V,VI)O_2^{+/2+}$, vor. Diese werden als Actinyl–Ionen bezeichnet. In dieser experimentellen Arbeit werden Sorptionsreaktionen von Neptunyl(V) mit der Mineralphase Calcit untersucht.

Die Calcitoberfläche wurde anhand von Zetapotential– und Oberflächen–Diffraktionsmessungen im Kontakt zu Lösungen unterschiedlichster Zusammensetzung charakterisiert. Adsorption von Calcium– und Carbonationen bestimmt das Zetapotential, während der pH–Wert nur einen geringen Einfluss hat. Es gibt keine Anzeichen für innersphärische Adsorption auf den Kristallebenen der Calcit–(104)–Oberfläche. Stattdessen wurden unter allen Bedingungen zwei geordnete Wasserschichten im Abstand von 2.35 ± 0.05 Å und 3.24 ± 0.06 Å über der Calcitoberfläche identifiziert. Daher wird ein Basic–Stern–Oberflächenkomplexierungsmodell für Calcit vorgeschlagen, das außer für Protonen und Hydroxidionen nur außersphärische Komplexierungsreaktionen in Betracht zieht.

Die Neptunyl(V)–Adsorption an der Calcitoberfläche wurde über einen weiten Bereich von Neptunylkonzentrationen und CO_2–Partialdrücken untersucht. Die Konzentrationsabhängigkeit der Adsorption zeigt den Verlauf von Freundlichisothermen. Die maximale Oberflächenbeladung entspricht einer zweiprozentigen Belegung der Oberflächenplätze. Die Adsorption ist pH–abhängig, maximale Adsorption wurde bei pH 8.3 beobachtet. Die pH–Abhängigkeit ist nur bis zu einer Gleichgewichtskonzentration von 10^{-5} M NpO_2^+ deutlich ausgeprägt. Bei pH 8.3 und einer hohen Oberflächenbeladung wurde mittels Tieftemperatur–EXAFS–Messungen (15K) ein bidentater innersphärischer Biscarbonatokomplex, der an Stufen auf der Calcit–(104)–Oberfläche gebunden ist, als häufigste Adsorptionspezies identifiziert.

In Kopräzipitationsexperimenten zeigt Neptunyl(V) eine hohe Affinität für den Einbau in die Calcitstruktur. EXAFS Messungen ergaben, dass im Calcit eingebautes Neptunyl in der äquatorialen Ebene von vier monodentat gebundenen Carbonationen koordiniert ist. Dementsprechend wird angenommen, dass Neptunyl in der Calcitstruktur ein Calcium und zwei Carbonationen substituiert. Wie der durch diesen Austausch erzeugte Ladungsüberschuss kompensiert wird ist nicht bekannt. Der Neptunyleinbau in die Calcitstruktur bremst das Calcitwachstum. Dies stimmt mit der Beobachtung überein, dass Neptunyl auf der Calcitoberfläche an Stufen adsorbiert, da insbesondere Stoffe, die an den beim Wachstum besonders reaktiven Stufen adsorbieren, das Wachstum schon in geringen Konzentrationen verlangsamen können. Aus Experimenten zur Adsorptionskinetik und aus Desorptionsexperimenten lässt sich ableiten, dass in begrenztem Umfang, auch wenn sich Calcit im Gleichgewicht mit der Lösung befindet, Neptunyl–Einbaureaktionen an der Calcitoberfläche stattfinden können. Dies konnte allerdings nicht spektroskopisch nachgewiesen werden.

Die Oberflächenreaktivität von Calcit, die Affinität von Neptunyl(V) für die Adsorption auf der Calcitoberfläche und die Möglichkeit des Einbaus von Neptunyl(V) in die Calcitstruktur, machen Calcit zu einer potentiell wichtigen Senke für Neptunyl(V) in der Geosphäre.

Meinen Eltern

Glauben Sie denen, die die Wahrheit suchen,
nicht denen, die sie gefunden haben.
nach *André Gide*, frz. Schriftsteller und Nobelpreisträger, 1869 – 1951

Contents

1. Introduction   **17**
    1.1. About the relevance of calcite 17
    1.2. Why study neptunyl(V)? 18
    1.3. Questions to answer 21

2. State of knowledge   **23**
    2.1. Chemical species in the aqueous $NpO2+$ – NaCl – CaCO3 system 23
       2.1.1. Aquatic species 24
       2.1.2. Solid phases 25
    2.2. The calcite surface 28
       2.2.1. Surface structure 28
       2.2.2. Calcite surface–speciation and –charge 32
    2.3. Calcite crystal growth 34
    2.4. Sorption of trace elements at the calcite surface 36
    2.5. Coprecipitation of trace elements with calcite 40
       2.5.1. Solid solution – aqueous solution equilibria 42
       2.5.2. Influence of trace element incorporation on crystal growth 45

3. Experimental details   **48**
    3.1. Analytical methods 48
       3.1.1. Scanning electron microscopy (SEM) 48
       3.1.2. X–ray photoelectron spectroscopy (XPS) 48
       3.1.3. Determination of the specific surface area by N2–BET 48

Contents

	3.1.4.	X-ray powder diffraction (powder XRD)	49
	3.1.5.	Inductively coupled plasma mass spectrometry (ICP–MS)	49
	3.1.6.	Inductively coupled plasma optical emission spectrometry (ICP–OES)	49
	3.1.7.	Liquid scintillation counting (LSC)	49
	3.1.8.	Error propagation calculations	50
3.2.	X-ray absorption fine structure (XAFS)	50
3.3.	pH measurements	54
3.4.	Calcite equilibration method	54
3.5.	Zetapotential measurements	55
	3.5.1.	Phase analyses light scattering (PALS)	55
	3.5.2.	Streaming potential measurements	57
3.6.	Surface diffraction	58
	3.6.1.	Basic principles of surface diffraction	58
	3.6.2.	Surface unit cell	61
	3.6.3.	Surface diffraction measurements	63
3.7.	Surface complexation modeling	67
3.8.	Adsorption experiments	71
	3.8.1.	The experimental setup and procedure	71
	3.8.2.	Low temperature (15 K) EXAFS measurements	72
3.9.	Coprecipitation experiments	73
	3.9.1.	Mixed flow reactor (MFR) experiments	73
	3.9.2.	EXAFS characterization of calcites from experiments Np1 – Np4 and U1	80
	3.9.3.	Near infrared (NIR) spectroscopy	82
	3.9.4.	Raman spectroscopy	82

4. Results and discussion   83
4.1.	The calcite surface	83
	4.1.1.	Zetapotential	83
	4.1.2.	Results from surface diffraction	86
	4.1.3.	Discussion of the surface structure	103
	4.1.4.	A new calcite surface complexation model	108
	4.1.5.	Inner–sphere adsorption at step sites?	116

Contents

    4.2. Neptunyl(V) adsorption to calcite 117
        4.2.1. Adsorption isotherms 117
        4.2.2. Adsorption kinetics, desorption, and irreversibility 118
        4.2.3. The adsorption complex structure 120
        4.2.4. Inclusion of neptunyl adsorption at calcite into the SCM? 127
    4.3. Neptunyl(V) coprecipitation with calcite 129
        4.3.1. MFR experiments 129
        4.3.2. Spectroscopic investigation of the incorporation sepcies 132
        4.3.3. Solid solution thermodynamics for neptunyl(V) doped calcite 141
    4.4. Comparison of neptunyl and uranyl interactions with calcite 144

5. Conclusions   **146**

6. Acknowledgment   **150**

Bibliography   **153**

A. ROD files   **166**

List of Figures

1.1. Left: Oxidation states of the minor actinides and right: Eh – pH diagram for neptunium in aqueous solution 20

2.1. Molar fractions (FNp) of the neptunyl complexes as a function of pH. . . . 26
2.2. Calcite structure and unit cells 27
2.3. SEM image of calcite and growth spirals at the calcite (104)–face. 29
2.4. Side and top view of the calcite (104)–face. 30
2.5. The critical radius for homogeneous nucleation 35
2.6. Growth rate determining processes 37
2.7. Neptunyl(V) adsorption at calcite. 39
2.8. PH dependence of neptunyl(V) adsorption at calcite. 40
2.9. Lippmann diagram for the calcite otavite solid solution. 45
2.10. Sites for trace element-adsorption 46

3.1. XAFS Experimental setup 51
3.2. XAFS regions 52
3.3. The experimental setup for calcite equilibration. 55
3.4. Principle of streaming potential measurements in the SurPASS elctrokinetic analyzer. 57
3.5. $|S_U|^2$ as a function of h. Peaks appear at integer values of h. Peak height is U^2 and peak width at half maximum is $2\pi/U$. 60
3.6. Calcite(104) surface unit cell. 62

List of Figures

3.7. Definitions of angles on a six (4+2) circle diffractometer and the diffractometer coordinate system. 63
3.8. Schematic drawing of the surface diffraction sample cell. 65
3.9. The Basic Stern model. 68
3.10. Setup for mixed flow reactor experiments 74
3.11. Detailed sketch of the MFR 75
3.12. Photographs of the experimental set up 76
3.13. SEM images of calcite crystals used as crystal seeds in MFR experiments. . 77

4.1. Results of electrophoretic PALS zetapotential measurements. 84
4.2. Streaming potential measurements on calcite in non equilibrium solutions. 85
4.3. Overview over CTR data of all datasets. 88
4.4. CTRs of dataset B. 91
4.5. Molecular structure of the calcite–water interface according to the result of the structure refinement of dataset B. 92
4.6. CTRs of dataset C. 93
4.7. Molecular structure of the calcite–water interface according to the result of the structure refinement of dataset C. 94
4.8. CTRs of dataset D. 95
4.9. Molecular structure of the calcite–water interface according to the result of the structure refinement of dataset D. 96
4.10. CTRs of dataset E. 97
4.11. Molecular structure of the calcite–water interface according to the result of the structure refinement of dataset E. 98
4.12. CTRs of dataset F. 99
4.13. Molecular structure of the calcite–water interface according to the result of the structure refinement of dataset F. 100
4.14. CTRs of dataset G. 101
4.15. Molecular structure of the calcite–water interface according to the result of the structure refinement of dataset G. 103
4.16. Electron density distribution, $\rho_{e^-}(z)$ (Å^{-3}), in the surface unit cell according to all six structures projected onto the $\mathbf{b_3}$ vector ($z \cdot \frac{\mathbf{b_3}}{|\mathbf{b_3}|}$ (Å)). 106

List of Figures

4.17. Comparison between literature data on calcite surface proton charge and corresponding SCM calculations. 110
4.18. Slip plane distances as a function of ionic strength. Calcite compared to various other surfaces. 112
4.19. Comparison of electrophoretic zetapotential data and SCM calculations. ... 113
4.20. Comparison of zetapotentials obtained by streaming potential measurements and SCM calculations. 114
4.21. Isotherms resulting from the adsorption experiments. Shown is surface load q (mol/g) versus solution concentration c (mol/L). 118
4.22. Experiments on adsorption kinetics. Shown is surface load q (mol/g) versus time (h). 119
4.23. k^3-weighted EXAFS data of samples Np-Y1, Np-O2, and for comparison an example for a spectrum of an incorporation species from a mixed flow reactor experiment. 120
4.24. a): k^3-weighted extracted EXAFS data and model spectrum of sample Np-O2. b): Fourier transform amplitude and imaginary part of the k^3-weighted EXAFS data and model spectrum. 121
4.25. Structure of the neptunyl triscarbonato complex with three bidentate bound carbonate ions. 123
4.26. Possible positions for bidentate adsorption of neptunyl at the calcite (104)–face. 124
4.27. Structure of the most likely neptunyl-calcite adsorption complex according to the results of the EXAFS data analyses. 126
4.28. AFM image of a freshly cleaved calcite (104)–face. 127
4.29. Neptunyl coprecipitation with calcite retards the calcite crystal growth. Plotted is steady state growth rate, R, versus steady state SI, SI_{out}. 131
4.30. Fourier transform amplitude and imaginary part of the k^2-weighted EXAFS data and model spectra. The original EXAFS data in k-space together with the corresponding fits are shown in the small diagrams above each Fourier transformed spectrum. 133
4.31. Structural model of a neptunyl-ion incorporated into the calcite host. Neptunium(V) is located on a Ca^{2+} site and the two axial neptunyl oxygens substitute for two adjacent carbonate ions. 136

List of Figures

4.32. Two equatorial coordination environments are possible for uranyl coprecipitated with calcite: one bidentate and three monodentate bound carbonate groups or two bidentate and only one monodentate bound carbonate groups. 138

4.33. Raman spectra of neptunyl doped and pure calcite. 139

4.34. Left diagram: NIR absorption spectra measured on three calcite single crystals from experiment Np11(SC). Right diagram: NIR absorption spectra of the neptunyl aquo–ion and the carbonato complexes. 140

4.35. Solid solution thermodynamic parameters can be derived from linear regression of $\ln D$ versus mole fraction X. 143

4.36. Neptunyl adsorption isotherms from this study compared to uranyl adsorption data from literature. 145

List of Tables

3.1. The A-matrix of the Basic Stern model for calcite. 70
3.2. Summary of experimental conditions for the MFR experiments. 81
4.1. Solution– and surface speciation (after the Pokrovsky and Schott SCM) at the conditions chosen for surface diffraction measurements. 87
4.2. Structural parameters obtained from fitting of dataset B. 92
4.3. Structural parameters obtained from fitting of dataset C. 94
4.4. Structural parameters obtained from fitting of dataset D. 96
4.5. Structural parameters obtained from fitting of dataset E. 98
4.6. Structural parameters obtained from fitting of dataset F. 100
4.7. Structural parameters obtained from fitting of dataset G. 102
4.8. Experimental conditions (CO_2 partial pressure and resulting pH) and Freundlich parameters resulting from the adsorption experiments. 117
4.9. Coordination numbers, N, distances, R (Å), and Debye-Waller factors, σ^2 (Å^2), from EXAFS data analyses. 122
4.10. Steady state conditions, pH_{out}, and SI_{out}, steady state growth rate, R, and partition coefficient, D, resulting from the MFR experiments 129
4.11. Results of the EXAFS analysis for coordination numbers, N, inter atomic distances, R, and Debye–Waller factors, σ^2. 134
4.12. The shift of spectroscopic parameters with subsequent carbonate complexation and calcite incorporation of NpO_2^+. 141

Chapter 1

Introduction

The disposal of high–level nuclear waste in deep geological formations poses major scientific and social challenges to be met in the next decades. One of the key issues is the long term safety of a waste repository over extended periods of time, up to 10^6 years. Over geological time spans it cannot be excluded that groundwater will infiltrate into a waste repository. It needs to be considered that permeating groundwater corrodes waste canisters and dissolves and thereby mobilizes radionuclides from the waste. Demonstrating the repository safety over geological time spans requires a sound understanding of the migration and retention behavior of radionuclides in the geosphere. In principle, sorption reactions between radionuclides and minerals control the retention. In recent years, various molecular-level sorption mechanisms have been identified: e.g. outer-sphere and inner-sphere adsorption, ion exchange, and structural incorporation (solid solution formation). The structural incorporation of trace elements into host minerals is not yet commonly considered in the safety analysis for repository systems, although these phenomena are quite common and extensively studied in natural systems. The main reason for this discrepancy is the lack of thermodynamic and kinetic data needed for the quantitative description of such processes.

1.1. About the relevance of calcite

Calcite with chemical formula $CaCO_3$ is the most common calcium carbonate polymorph. Other polymorphs like the orthorhombic aragonite and the hexagonal vaterite are thermodynamically not stable at standard conditions. Aragonite plays an important role in

1. Introduction

mollusk shells, in hydrothermal formations, and in metamorphous rocks. Vaterite can form as precursor phase of calcite from highly supersaturated solutions. Calcite crystallizes in the hexagonal space group, $R\bar{3}c$. It has a density of $2.71\,\text{g}/\text{cm}^3$ and defines the mohs' hardness 3 [1].

Calcite is one of the most common minerals on earth. About 4% of earth's crust consist of calcite [2]. Most of the carbonate minerals are found in sedimentary rocks. Limestones are formed mainly of biogenic carbonates but under special conditions carbonates also precipitate from natural waters. Argillites can contain large amounts of carbonate minerals and there is a fluent transition from limestone over clayey limstone to clay. In magmatic rocks carbonates are found in carbonatites. Carbonate minerals and the ions of carbonic acid define to a high degree the geochemical milieu of groundwater. Argillites are in addition to saltrocks and granite, proposed as possible geological formation for nuclear waste disposal, and contain 10 % to 40 % calcite [3]. Calcite is a possible alteration product of concrete based materials, part of the technical barrier in the multi-barrier system meant to prevent radionuclide release from the waste repository. Calcite is a common fracture filling material in some granitic rocks considered as host rock formation for nuclear waste repositories in Scandinavia [4]. Numerous studies show that trace elements adsorb at the calcite surface and can be incorporated into the calcite crystal structure [5, 6, 7]. Therefore interactions between calcite and actinides will play an important role in actinide migration in the near and far field around nuclear waste repositories. Its reactivity makes this mineral a potentially important sink for radionuclides, as well as for other heavy metal contaminants in the geosphere.

1.2. Why study neptunyl(V)?

Neptunium with the atomic number 93 is the first transuranium element. It belongs to the minor actinides. With a half-life of $2,144 \cdot 10^6$ years neptunium-237 is the longest-lived neptunium isotope. In nature, only trace amounts of neptunium exist in uranium ores. Its abundance in earth's crust can be calculated from the natural decay chains to be $4 \cdot 10^{-17}$ %. The far larger part of existing neptunium is produced artificially in nuclear reactors. Therefore, it is often called an artificial element. In German nuclear power plants driven with moderately enriched uranium fuel several kilogramms of neptunium-237 are produced per year [8]. Higher amounts of neptunium-237 are formed as an intermediate

1. Introduction

product during plutonium-238 production. In conventional nuclear power plants driven by moderately enriched uranium fuel, neptunium-237 forms by two reactions [9]. 70 % is formed according to the reaction:

$$^{238}_{92}U \quad (n,2n) \quad ^{237}_{92}U \quad \xrightarrow{\beta^-} \quad ^{237}_{93}Np \tag{1.1}$$

the remaining 30 % form according to the reaction:

$$^{235}_{92}U \quad (n,\gamma) \quad ^{236}_{92}U \quad (n,\gamma) \quad ^{237}_{92}U \quad \xrightarrow{\beta^-} \quad ^{237}_{93}Np \tag{1.2}$$

Positron decay of plutonium-237 and alpha decay of americium-241 also lead to neptunium-237 [10].

Calculations show that after long periods of time ($\sim$800,000 years) Np-237 becomes one of the major sources of radiotoxicity in spent fuel because of its long half-life and because of its formation in the decay chain of Pu-241 and Am-241 [8].

Dozol and Hagemann (1993) [11] and Clark et al.(1995) [12] emphasize the importance of neptunium-237 when considering longterm performance assessment of nuclear waste disposals.

> "In particular the evaluation of the transport properties of ^{237}Np is of great interest due to the high ingestion radiotoxicity of that isotope." [11]
> "Neptunium is the most problematic actinide element with respect to environmental migration because its solubility under typical groundwater conditions is expected to be high enough to be of radiological concern and its sorption on common minerals is expected to be relatively low." [12]

The latter is especially true for oxidizing or redox–neutral groundwater conditions where pentavalent neptunium, neptunyl(V), is dominant. In aqueous solution neptunium can exist in the oxidation states III, IV, V, and VI (see Figure 1.1). Under extremely oxidizing conditions even neptunium(VII) has been observed. Neptunium(III) and neptunium(IV) exist as free ions in solution coordinated by nine water molecules while neptunium(V) and neptunium(VI) form near linear trans–dioxo–cations, the so called neptunyl ions, that show a fivefold equatorial water coordination [13]. In natural environments only the tetra– and pentavalent oxidation states are relevant. Similarly as for tetravalent uranium and uranyl(VI), tetravalent neptunium is less soluble and sorbs more strongly to mineral surfaces than neptunyl(V). The expected high mobility of neptunyl(V) makes it a key uncertainty for the long term safety of nuclear waste disposal systems.

1. Introduction

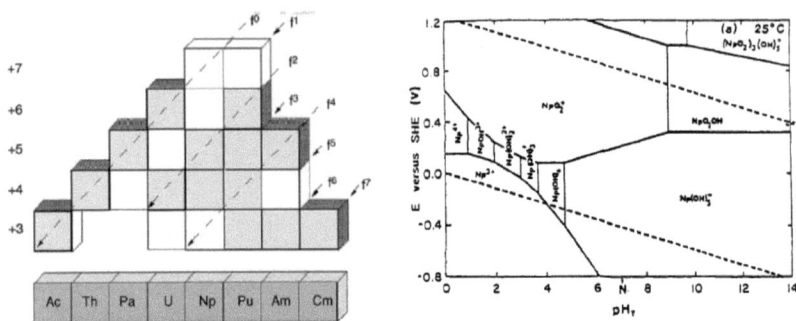

Fig. 1.1: Left Figure: Known oxidation states of the minor actinide elements (indicated at left) and their f-electron configuration (indicated at right). The oxidation states stable in natural environments are highlighted [14]. Right Figure: Eh–pH diagram for neptunium in aqueous solution [11].

Often lanthanides are used as non–radioactive homologous to study the chemical properties of trivalent actinides. Zhong and Mucci (1995) reported partitioning coefficients between rare earth elements and calcite [15]. Structural investigations by Elzinga et al. (2002) [16] followed. Curti et al. [17] used an advanced thermodynamic "inverse" modeling approach to model the Eu/calcite solid solution–aqueous solution (SSAS) system. Recent studies show that the trivalent actinides Am^{3+} and Cm^{3+} can be incorporated into calcite, too [18, 19].

For uranium, neptunium, and plutonium similar approaches are not strictly applicable due to the fact that these elements exist in various oxidation states including the linear actinyl ions, $An(V,VI)O_2^{(+,2+)}$ under environmentally relevant conditions. The chemistry of the actinyl ions is unique among all the elements of the periodic table and non–radioactive homologues do not exist.

Among the actinide elements, uranium is the most researched and its interaction with calcite has been subject of numerous studies. Uranium(IV) in calcite was studied by Sturchio et al.(1998) [20]. Elzinga et al. (2004) studied the adsorption of hexavalent uranyl at the calcite water interface [21]. Reeder et al. reported the incorporation of uranyl into artificially synthesized calcite [22, 23, 24]. Kelly et al. investigated the structural environment of uranyl in natural calcite [25, 26].

1. Introduction

Before our study [27] no publications about coprecipitation of neptunyl(V) with calcite existed, but neptunium(V) was known to adsorb more strongly to calcite than to most other minerals [28, 29].

1.3. Questions to answer

This work especially addresses the structural incorporation of neptunyl(V) into calcite. Structural incorporation of radionuclides into mineral phases leading to solid solution formation in low temperature aquatic environments involves complex interactions at the mineral solution interface. A first reaction step includes the adsorption of the metal cation at the mineral surface. Adsorption is expected to be influenced by the mineral surface speciation, –structure, and –nanotopography. To accomplish structural incorporation there must be a thermodynamic driving force for the incorporation reaction. The aim of this study is to acquire a mechanistic understanding of the incorporation processes at a molecular scale, quantify the sorption reactions, and explore the possibilities to describe observed adsorption and incorporation reactions by thermodynamic models.

These goals define the questions that are addressed in this experimental study:

- What defines the reactivity of the calcite surface? How do surface structure, charge, and speciation vary as a function chemical conditions?

- How much neptunyl(V) adsorbs at the calcite surface and what is the nature of the adsorption complexes?

- Can neptunyl(V) ions be incorporated into the calcite structure, and if yes, to what extent? Is the resulting solid solution thermodynamically stable?

The following experimental approach has been chosen to tackle these questions:

- The calcite–water interface is investigated by means of zetapotential and surface diffraction measurements. These studies provide the basis for the development of a new surface complexation model for calcite.

- Neptunyl(V) adsorption at the calcite surface is quantified over an extended range of neptunyl concentrations and CO_2 partial pressures. The structure of a neptunyl adsorption complex at the calcite water interface is characterized by x–ray absorption spectroscopy.

1. Introduction

- Coprecipitation experiments are used to quantify the incorporation of neptunyl into calcite. The structure of the incorporation species is characterized by x–ray absorption spectroscopy. Additional spectroscopic techniques are applied to validate the XAS results. Thermodynamic parameters for neptunyl doped calcite are discussed.

Chapter 2

State of knowledge

2.1. Chemical species in the aqueous NpO_2^+ – NaCl – $CaCO_3$ system

A multitude of chemical species exists in the aqueous NpO_2^+ – NaCl – $CaCO_3$ system. The aquatic speciation as well as the solubility of solid phases at standard conditions has been studied extensively over the past decades. Thermodynamic data is available for speciation and solubility calculations at standard conditions for many solution species and phases. However, there is still some uncertainty related to the completeness of the thermodynamic data for neptunium.

Recently, in studies addressing uranium and plutonium solubility and complexation in calcium containing systems new ternary calcium – uranium/plutonium aqueous complexes have been discovered that caused increased solubility [30, 31]. Similar effects have not yet been studied systematically for neptunium. The most relevant neptunium redox–reaction, the transition:

$$Np(V)O_2^+ + 4\,H^+ + e^- \rightleftharpoons Np^{4+} + 2\,H_2O$$

is not yet completely understood. This redox reaction involves not only electron transfer but also protonation and dehydration reactions. It is a current topic of research if this reaction can take place in solution or if intermediate solid or colloidal phases or reactive surfaces are necessary. Described below are the reactions that were considered in thermodynamic speciation and supersaturation calculations in this work. If not otherwise indicated equilibrium constants are taken from the Nagra/PSI thermodynamic database [32]. Thermodynamic model calculations are performed using either PhreeqC [33] or Ecosat [34]. In these software packages the Davies' equation is used as default for activity corrections.

2.1.1. Aquatic species

Carbonate

The total amount of dissolved inorganic carbon in solution is directly related to the CO_2 partial pressure in the surrounding gas phase and solution pH. In thermodynamic model calculations carbonic acid and dissolved $CO_{2(aq)}$ concentrations are usually not considered separately, but the bicarbonate concentration is directly related to the CO_2 partial pressure in the equilibrium gas phase. $CO_{2(g)}$, bicarbonate, and carbonate are related by the following reactions:

$$CO_{2(g)} + H_2O \rightleftharpoons HCO_3^- + H^+ \quad (\log_{10} K = -7.82)$$
$$HCO_3^- \rightleftharpoons CO_3^{2-} + H^+ \quad (\log_{10} K = -10.33)$$

Calcium

The alkaline earth element calcium exists in aqueous solutions as the divalent Ca^{2+} cation. In the investigated system Ca^{2+}, as well as $CaHCO_3^-$, $CaCO_{3(aq)}$, and $CaOH^+$ species, can form. The calcium solution speciation is defined by the reactions:

$$Ca^{2+} + HCO^- \rightleftharpoons CaHCO_3^+ \quad (\log_{10} K = 1.11)$$
$$Ca^{2+} + HCO^- \rightleftharpoons CaCO_{3(aq)} + H^+ \quad (\log_{10} K = -7.10)$$
$$Ca^{2+} + H_2O \rightleftharpoons CaOH^+ + H^+ \quad (\log_{10} K = -12.78)$$

Neptunium

Neptunium exists in the the oxidation states zero and three to seven, but only tetra– and pentavalent neptunium are stable under environmentally relevant Eh-pH conditions. Pentavalent neptunium forms the neptunyl ion, NpO_2^+, in aqueous solutions. Neptunyl is to be expected in oxygen rich natural waters, while under reducing conditions Np(IV) will form [12, 11]. As described above (2.1) it is not yet well understood under which conditions the transformation from NpO_2^+ to Np^{4+} takes place. For the experiments in this study pentavalent neptunium is used and it is assumed that no neptunium redox reactions take place. This assumption has frequently been verified by spectroscopic investigations.

Due to its low effective charge the complex formation constants for NpO_2^+ are low compared to the other actinide ions. The strength of complexing agents decreases in the sequence $CO_3^{2-} > OH^- > F^- > PO_4^{3-} > SO_4^{2-} > Cl^-, NO_3^-$. Neptunyl does not form complexes with Cl^- and NO_3^-. Known neptunyl-carbonato, and -hydroxo complexes form according to the reactions:

$$NpO_2^+ + H_2O \rightleftharpoons [NpO_2OH] + H^+ \quad (\log_{10} K = -11.3)$$
$$NpO_2^+ + 2H_2O \rightleftharpoons [NpO_2(OH)_2]^- + 2H^+ \quad (\log_{10} K = -23.6)$$
$$NpO_2^+ + CO_3^{2-} \rightleftharpoons [NpO_2CO_3]^- \quad (\log_{10} K = 4.96)$$
$$NpO_2^+ + 2CO_3^{2-} \rightleftharpoons [NpO_2(CO_3)_2]^{3-} \quad (\log_{10} K = 6.53)$$
$$NpO_2^+ + 3CO_3^{2-} \rightleftharpoons [NpO_2(CO_3)_3]^{5-} \quad (\log_{10} K = 5.50)$$
$$NpO_2^+ + 2CO_3^{2-} + H_2O \rightleftharpoons [NpO_2(CO_3)_2OH]^{4-} + H^+ \quad (\log_{10} K = -5.30)$$

A speciation plot considering hydroxo- and carbonato complexes is shown in Figure 2.1. Up to pH 8.5 the free neptunyl ion dominates the aquatic speciation Above that the neptunyl monocarbonato complex dominates the solution speciation. At pH above 12.5 the ternary hydroxo–carbonato– and the dihydroxo–complexes start to play an important role.

2.1.2. Solid phases

Calcite is the most stable of the calcium carbonate polymorphs. The decreasing stability in the series calcite, aragonite, vaterite is reflected in their increasing solubility product (K_{SP}). At high pH it is important to consider the calcium hydroxide compound portlandite ($Ca(OH)_2$) in addition to the calcium carbonate polymorphs:

$$\log_{10} K_{SP} \text{ (Calcite)} = -8.48$$
$$\log_{10} K_{SP} \text{ (Aragonite)} = -8.34$$
$$\log_{10} K_{SP} \text{ (Vaterite)} = -7.91$$
$$\log_{10} K_{SP} \text{ (Portlandite)} = -5.20$$

It is not always the thermodynamically most stable solid that forms from a supersatu-

2. State of knowledge

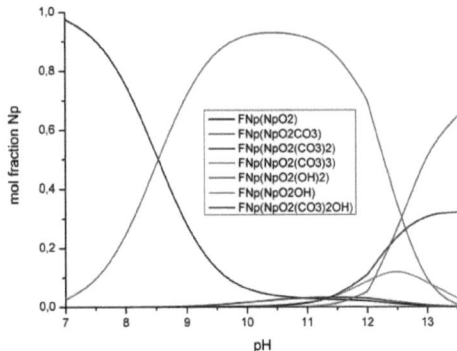

Fig. 2.1: Molar fractions (FNp) of the neptunyl complexes as a function of pH. Total neptunium concentration 1 μM, total inorganic carbon (TIC) concentration 0.8 mM, ionic strength 0.01 M NaCl.

rated solution. Non–equilibrium systems can equilibrate by step wise loss of Gibb's free energy. There is no general theoretical description of this phenomenon, but the most likely explanation is that the kinetics of the phase transitions, the relative rates of crystal nucleation and growth, decide which phase forms [35]. This has to be kept in mind when dealing with $CaCO_3$. From highly supersaturated calcium carbonate solutions metastable vaterite may precipitate as a precursor phase and later transform to calcite [36].

Calcite crystallizes in the rhombohedral space group $R\bar{3}c$. Aragonite is orthorhombic and shows the space group symmetry $Pmcn$. Vaterite is a hexagonal mineral with space group symmetry $P6_3/mmc$. Figure 2.2 shows the crystal structure of calcite. Depicted in Figure 2.2 are three different possible unit cells for the calcite structure. The pink one is probably the most descriptive unit cell. It can be imagined as a cubic face centered structure with one shortened space diagonal. The faces of this unit cell are the ones along which calcite shows perfect cleavage. These six faces are transformed into each other by the $\bar{3}$–symmetry. Calcite crystals, microscopic powders, as well as big crystals, often show a rhombohedral habit just like magnifications of this unit cell. The green lines in Figure 2.2 show another rhombohedral unit cell, which is not especially relevant. The yellow unit cell is the hexagonal unit cell, which is most commonly used in calcite crystallography. According to this

2. State of knowledge

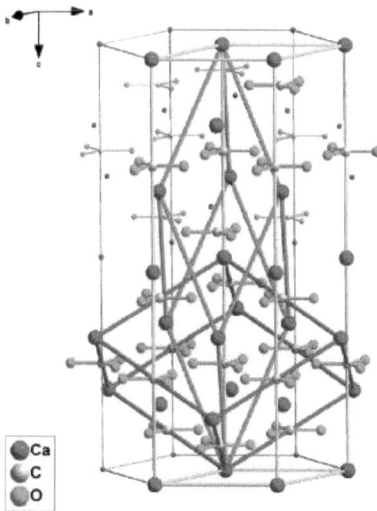

Fig. 2.2: Three possible unit cells for the calcite structure [38].

unit cell the planes of perfect cleavage are defined by the reciprocal (104)–vector. The cell dimensions differ slightly for different calcite crystals depending e.g. on the purity, the conditions during crystal growth, temperature, and pressure. Cell parameters used in this study are: $b_1 = b_2 = 4.988$ Å , $b_3 = 17.061$ Å, $\alpha = \beta = 90°$, and $\gamma = 120°$ [37].

Compared to the calcium carbonate phases much less is known about solid neptunium phases. Thermodynamic data is available for three phases that could appear in the investigated system. Two of them show precursor phases that need to be considered when dealing with neptunium solubilities. The solids marked with "fr" in the list below("fr" stands for freshly precipitated) are precursor phases that transform upon aging or annealing to the corresponding ones marked "ag" (aged). As for vaterite and calcite the precursors form faster than the stable phases and exhibit higher solubilities. The solubilities of the three neptunyl phases differ by orders of magnitude:

2. State of knowledge

$\log_{10} K_{SP} (NpO_2(OH)(\text{am, fr}))$ = 5.30
$\log_{10} K_{SP} (NpO_2(OH)(\text{am, ag}))$ = 4.70
$\log_{10} K_{SP} (NaNpO_2CO_3 \cdot 3.5H_2O(\text{s, fr}))$ = -11.00 [39]
$\log_{10} K_{SP} (NaNpO_2CO_3(\text{s, ag}))$ = -11.66
$\log_{10} K_{SP} (Na_3NpO_2(CO_3)_2(\text{s}))$ = -14.70

Neptunyl hydroxide has a very high solubility compared to the carbonate phases. The solid phase with the lowest solubility product is $Na_3NpO_2(CO_3)_2$. However, it is only relevant at elevated sodium concentrations. Calculations show that $NaNpO_2CO_3 \cdot 3.5H_2O$ (s, fr) is the solubility limiting phase for the experiments in this study. Volkov et al. (1979) [40] have studied the structures of pentavalent neptunium carbonate phases with stoichiometry: $MNpO_2CO_3$, with "M" being various alkaline metals. They report that neptunyl and carbonate ions form hexagonal or orthogonal layers depending on the size of the alkaline metal interlayer cations. Interlayer sodium can be exchanged for calcium forming a solid of the composition: $Ca_{0.5}NpO_2CO_3$. This mineral phase could be relevant for this study. There is however no solubility data available.

2.2. The calcite surface

Mineral–ion interactions and crystal growth take place at the solid liquid interface. Therefore, when studying sorption, crystal growth, and coprecipitation phenomena it is most important to know about the structure, charge, and speciation of the mineral surface.

2.2.1. Surface structure

Natural calcite displays a wide variety of morphologies and crystals sometimes display combinations of scalenohedrons, prisms and rhombohedrons. Surfaces are composed of the crystallographic (001), (110), (012), and (104)–faces, but the (104)–face is by far the most abundant. An scanning electron microscopy (SEM) image of the commercial calcite used for many of the experiments in this study is shown on the left hand side in Figure 2.3. Even though the crystallites are agglomerated the rhombohedral shapes caused by the dominance of the (104) face are obvious. However, what looks like broken off corners of the rhombohedra on a few crystallites might be (001), (110), or (012) faces. The abundance of different crystallographic faces can be satisfactory explained by the Periodic Bond Chain

2. State of knowledge

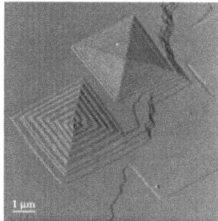

Fig. 2.3: The left image shows an SEM image of commercial calcite powder and the right image shows growth spirals at the calcite (104)–face growing along steps parallel to the [$\bar{4}$41] and the [48$\bar{1}$] directions [43].

(PBC) model of Hartman and Perdok (1955) [41]. Paquette and Reeder (1995) identified three nonequivalent PBCs in the calcite structure along the [$\bar{4}$41], the [2$\bar{2}$1], and the [010] direction [42]. (Strictly it is not the [2$\bar{2}$1] direction that lies in the (104)–plane but the symmetry equivalent [42$\bar{1}$] direction. In the following the latter vector will be used.) The (104)–face contains all three of them. The [$\bar{4}$41] PBC is the straightest and is therefore considered the most stable PBC. It is present twice on the calcite (104)–face in non–parallel directions. This explains the high stability of this crystal face and is the reason for the high abundance of this face on crystals grown from solution. It also explains the perfect cleavage along this plane. In the lower part of Figure 2.4 the four PBCs are shown as green bonds between the ions. Due to the high abundance of the (104)–face at crystals grown from solution it seems reasonable to assume that the error made by taking the properties of the (104)–face as representative for the whole calcite surface is small. The perfect cleavage along the (104)–plane is another more practical reason why it is the by far most studied calcite face. It is simply the one easiest to prepare.

The presence of more than two PBCs in a crystal face also has some other interesting implications. According to the PBC model such faces are considered flat, as opposed to stepped (one PBC in the plane) or kinked (no PBC). Flat crystal faces do not grow by attachment of single molecules to the flat surface. Instead growth proceeds along steps on the surface which are parallel to the most stable PBCs. For calcite that means parallel to the [$\bar{4}$41] direction or the symmetry equivalent [48$\bar{1}$] direction. This is exactly what can be observed in in situ AFM crystal growth experiments [44, 43, 45]. Theoretically

2. State of knowledge

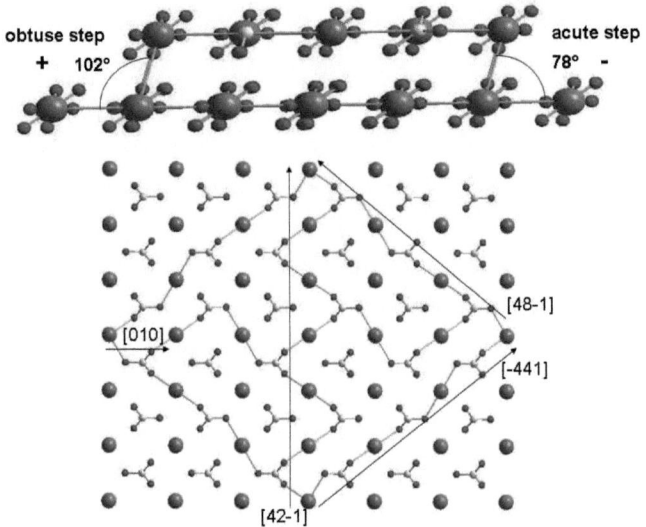

Fig. 2.4: Top: Side view of the calcite (104)–face along the $[\bar{4}41]$ direction showing an obtuse and an acute step. Bottom: Top view of the calcite (104)–face. Green bonds mark the PBCs. The arrows show the corresponding crystallographic vectors. (oxygen: red, carbon: gray, and calcium: cyan)

this also implies that the crystals would stop growing as soon as all steps have reached the crystal edges. Indeed screw dislocations are necessary as sources of steps to keep the crystal growing and to explain observed growth rates at low supersaturations [46]. The right part if Figure 2.3 shows growth spirals at the calcite (104)–face growing along steps parallel to the $[\bar{4}41]$ and the $[48\bar{1}]$ directions.

In the top part of Figure 2.4 it is shown that steps along the $[\bar{4}41]$ direction on either side of an "island" on the surface, as it could for example, be created by a 2D growth nucleus, do not have the same geometry. There is an obtuse edge with an angle of 102° towards the surface and an acute edge with an angle of 78° towards the surface. The same step geometries exist along the $[48\bar{1}]$ direction. For each of the two different step geometries (the acute and the obtuse step) two different kink sites exist during growth. These four different kink sites, which are considered the most reactive sites during crystal growth, ad-

2. State of knowledge

sorption, and coprecipitation, do not differ regarding the coordination numbers and bond distances to the nearest and second nearest carbonate and calcium neighbors. Only the angular positions of the neighboring atoms are different. Nevertheless many studies have shown that they behave differently regarding crystal growth and trace element incorporation [42, 44, 43, 45, 47] resulting in anisotropic growth features and intrasectoral zoning effects.

The atomic structure of the calcite(104)–face corresponding to the bulk calcite structure is shown in Figure 2.4. The rectangular arrangement of the calcium ions on the (104)–face can be seen in the lower part of the Figure. Distances between the calcium ions are 4.988 Å in the [010] direction and 4.047 Å in the [42$\bar{1}$] direction. The carbonate groups located with the carbon atom exactly in the middle of the four calcium neighbors show an alternating pattern along the [42$\bar{1}$] direction, depicting the glide plane symmetry along this axis. The carbonate groups comprise an angle of 44.7° with the surface. The calcium to carbonate ratio at the surface is 1:1. The crystallographic density of calcium and carbonate sites on the calcite (104)–face is 4.95 nm^{-2} or $8.22 \cdot 10^{-6}$ mol/m^2.

The structure of the calcite(104)–water interface has been subject of various surface diffraction studies. Fenter et al. (2000) [48] have studied the calcite(104)–water interface at pH 6.8, 8.3, and 12.1 under atmospheric CO_2 by means of x-ray reflectivity. They report on a full monolayer of a water or hydroxyl at 2.50±0.12 Å above the surface. The position of the surface calcium atom defines as in most other studies the location of the surface. The surface carbonate groups are tilted towards the surface. The reflectivity data is best modeled without including any calcium or carbonate inner–sphere complexes. The uncertainties for the surface site occupancies range from 4 % to 9 %. This implies that surface diffraction is generally very sensitive to inner–sphere adsorption. A surface diffraction study analyzing the full 3D structure of the calcite–water interface at pH 8.3 in equilibrium with atmospheric CO_2 has been published by Geissbühler et al. (2004) [49]. Their results show that there are two well ordered water layers above the surface the first 2.30 ± 0.1 Å above the surface and the second one 3.45 ± 0.2 Å above the surface. Ions in the first two monolayers of the surface slightly relax from their bulk positions. The surface carbonate ions are tilted towards the surface by 11.3°. The distance between the surface calcium and the closest water molecule is 2.97 ± 0.12 Å.

Magdans et al. (2007) and (2005) [50, 51] have also studied the calcite(104)–water interface

2. State of knowledge

by means of surface diffraction. As contact solution they use destilled water. Their results differ slightly from those of Geissbühler et al. (2004). Especially the first water layer is in their case much closer to the surface (1.9 ±0.1 Å) and the distance between surface calcium and the water molecules of the first water layer is only 2.4 ±0.1 Å.

A huge amount of literature is also available on theoretical studies about the calcite(104)–water interface at various levels of theory [52, 53, 54, 55, 56, 57]. Some of the recent theoretical results are in good agreement with the experimental x-ray surface diffraction results. For example, Kerisit and Parker (2004) [55] studied free energy and density profiles of water and metal sorption above the calcite (104)–surface by molecular dynamics simulations. They find three minima in the free energy of water adsorption and correspondingly three maxima in water density at 2.2 Å, 3.2 Å, and 5.0 Å above the surface. While theory seems to have developed methods to reliably predict the structure of the interface and the energetics of some surface reactions like adsorption of some ions at the surface or dissolution of ions from steps or from the flat surface, many of the reactions that are most relevant to predict the surface charge and speciation cannot yet be tackled by theoretical methods.

2.2.2. Calcite surface–speciation and –charge

Surface charge and zetapotential of calcite have been studied by many researchers over the past decades. Depending on the calcite used and the conditions under which the measurements are conducted the values for the isoelectric point (IEP) vary between 7 and 11 [58]. Values for the Point of Zero Charge (PZC) can be found ranging from 8 to 9.5 [59]. For oxide minerals the PZC and the proton surface charge as a function of pH is determined by acid – base titrations. For calcite a similar approach is hardly applicable as fast dissolution kinetics and the buffering effect of the carbonate ions in solutions adulterate the results. Therefore the exact mechanisms that determine the surface charge and potential of calcite are not yet well understood. Van Cappellen et al. (1993) [60] dealt with that problem by deriving a surface complexation model (SCM) for calcite in analogy to previously published titration data on the less soluble mineral phases siderite ($FeCO_3$) and rhodochrosite ($MnCO_3$) [61] based on thermodynamic constants for calcium and carbonate hydrolysis and protonation reactions of the solution species. Pokrovsky et al. (2000) have revised the resulting Constant Capacitance Model based on diffuse reflectance infrared (DRIFT) spectroscopic results. They assign features in the DRIFT spectra measured on wet calcite

2. State of knowledge

pastes at 3400 cm^{-1} and 1420 cm^{-1} to surface hydroxyl– and surface carbonate groups. In low pH solutions they also find an increase in the peak around 3400 cm^{-1} and explain it by an increasing number of hydrated calcium outer–sphere complexes [62]. Pokrovsky and Schott have published a newer version of the same model in 2002 in comparison with analogous models for many other divalent metal carbonates with slightly changed constants. These are surface reactions considered in this model and the related intrinsic stability constants. Surface species are designated as ">":

$$>CO_3H \quad\rightleftharpoons\quad >CO_3^- + H^+ \quad (\log_{10} K = -5.1)$$
$$>CO_3H + Ca^{2+} \quad\rightleftharpoons\quad >CO_3Ca^+ + H^+ \quad (\log_{10} K = -1.7)$$
$$>CaOH \quad\rightleftharpoons\quad >CaO^- + H^+ \quad (\log_{10} K = -12.00)$$
$$>CaOH + H^+ \quad\rightleftharpoons\quad >CaOH_2^+ \quad (\log_{10} K = 11.85)$$
$$>CaOH + CO_3^{2-} + 2H^+ \quad\rightleftharpoons\quad >CaHCO_3^0 + H_2O \quad (\log_{10} K = 23.50)$$
$$>CaOH + CO_3^{2-} + H^+ \quad\rightleftharpoons\quad >CaCO_3^- + H_2O \quad (\log_{10} K = 17.1)$$

The electric double layer capacitance in 0.01 M NaCl solution used in this model is 17 F/m^2. As surface site density they use the crystallographic value for the (104)–face of 8.22 mol/m^2. Recently a simplification of this model has been proposed. As the protonated and deprotonated species' abundance at the >Ca and the >CO$_3$ site vary in a very similar way with changing pH they can be put together to one generic >CaCO$_3$ site without loosing much of the model accuracy [63].

The major draw back of the constant capacitance models for the divalent metal carbonate minerals are the unrealisticly high capacitance values that are needed to link the high surface charge and the relatively low zetapotentials. The capacitance (C_1) describes the relation between surface charge (σ_0) and the surface potential (ψ_0): $\sigma_0 = C_1\psi_0$. In principle the capacitance of a parallel–plate capacitor is given by $C_1 = \frac{\epsilon_0 \cdot \epsilon_r}{x}$, with ϵ_0 being the permittivity of free space ($\epsilon_0 = 8.8542 \cdot 10^{-12}$ F/m), ϵ_r being the relative permittivity of the medium between the capacitor plates. x is the distance between the capacitor plates. Even assuming the high relative permittivity of liquid water for the electric double layer ($\epsilon_r = 78.5$) the thickness of the electric double layer would be 0.4 Å. Based on this low value Charlet et al. (1990) [61] explain the high surface charges found for siderite and rhodochrosite as an indication for a completely collapsed electric double layer.

Stipp (1999) [64] has studied the calcite surface by means of x-ray photoelectron spectroscopy (XPS) and concluded that there is a hydrolysis layer at the surface. Based on

2. State of knowledge

zetapotential data by Foxall et al. (1979) [65] measured as a function of calcium concentration and her observations, she suggests that the potential determining ions are Ca^{2+} and the different carbonate species, and that the sorption of H^+ and OH^- to the surface only have a minor influence on the surface charge. H^+ and OH^- are in so far important as pH controls the carbonate speciation. Correspondingly, she has proposed an enhanced electrical double layer model to describe the potential above the calcite surface. In order to be consistent with the XPS observations a hydrolysis layer is included into the Stern layer between the bulk solid and the potential determining ions in an outer–sphere layer. A most recent theoretical study has shown that H^+ and OH^- ions next to each other at the surface are not stable and would immediately associate and form H_2O [57].

Wolthers et al. (2008) [58] have been the first that developed a more sophisticated calcite SCM applying the Charge Distribution MUlti Site Comlexation, CD–MUSIC, modeling approach [66]. In this model an inner–sphere and an outer–sphere adsorption layer exist. The charge of adsorbing ions is spread over the two layers. This model offers a far more realistic representation of the processes at the calcite solution interface. Still the problem of the high capacitance values has not been overcome.

2.3. Calcite crystal growth

Definition of saturation expressions

In principle a crystal can grow if it is in contact with a supersaturated solution. Different ways exist to express the saturation state of a solution. For a mineral with composition $A_a B_b$ the solubility product, K_{SP}, is defined as:

$$K_{SP} = [A]_{eq}^a [B]_{eq}^b \tag{2.1}$$

with the equilibrium activities of the ions $[A]_{eq}$ and $[B]_{eq}$. If

$$S = \frac{[A]^a [B]^b}{[A]_{eq}^a [B]_{eq}^b} > 1, \tag{2.2}$$

that means the actual activities $[A]$ and $[B]$ are higher than the equilibrium activities, the solution is supersaturated. S is called saturation. Often it is more convenient to use the saturation index, SI, in stead of S. It is defined as:

$$SI = \log_{10} S. \tag{2.3}$$

2. State of knowledge

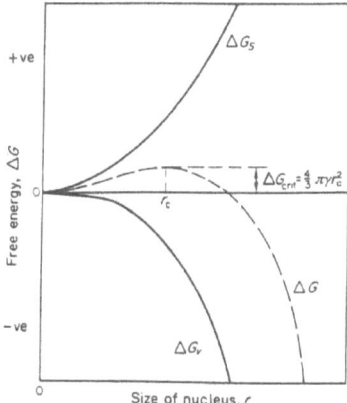

Fig. 2.5: The critical radius, r_c, a crystal nucleus has to overcome to become stable can be displayed as a maximum in the overall excess free energy, ΔG, of the particle.

For undersaturated solutions $SI < 0$, for equilibrium solutions $SI = 0$, and for supersaturated solutions $SI > 0$.

Homogeneous nucleation

For very small particles the solubility increases with decreasing particle size. Therefore the particle size has to exceed a distinct critical radius for crystal nuclei to become stable. For spherical nuclei with radius, r, this critical radius can be calculated by considering the overall excess free energy of the particle (ΔG), the sum of surface excess free energy (ΔG_S), and the volume excess free energy (ΔG_V). The surface energy is positive and proportional to r^2, while the volume energy is negative and proportional to r^3. As shown in Figure 2.5 the overall excess free energy of the particle reaches a maximum for a distinct radius, the critical radius, r_c. For $r > r_c$ it is energetically preferable for the particle to grow and for $r < r_c$ redissolution is preferred. In solutions with low supersaturation the time for the formation of a growth nucleus with a size, $r > r_c$, becomes practically infinite. In such solutions crystal growth can only take place at available surfaces. They are called metastable solutions. Solutions can be called metastable with respect to calcite if SI(calcite) < 1 [67].

2. State of knowledge

Surface controlled growth, kinetics and mechanisms

In agitated supersaturated solutions a thin film of laminar flowing water develops above the surface of suspended crystals. In this film a concentration gradient is established between the turbulent mixed bulk solution and the mineral surface, where the concentration is diminished by incorporation of ions onto the mineral. Ions move through this film perpendicular to the surface by diffusion. The thickness of this diffusion–film depends on the relative velocity between the solution and the mineral particle. In analogy to the Stokes' settling rate this velocity depends on particle size. Above a critical particle size the relative velocity between solution and crystal can be controlled by the agitation velocity in the solution and the influence of diffusion on the reaction rate can be minimized by appropriate agitation. This phenomenon is known as boundary layer effect. The critical particle size is around 5 to 10 μm [36].

Diffusion controlled growth processes exhibit a linear dependence between the growth rate and supersaturation. For calcite the influence of surface processes on the growth rate predominates over the influence of diffusion in the whole metastable supersaturation range [36]. Accordingly, Tai (1999) [68] have not found any dependence of the calcite growth rate on the solution velocity above the mineral surface.

As shown in Figure 2.6 for low supersaturations the growth rate is surface controlled. Different growth processes run parallel and always the fastest one is rate determining. For very low supersaturation conditions spiral growth determines the rate and with increasing supersaturations it is overtaken by growth from surface nuclei. Then at some point the transport of ions to the surface that precedes both of these processes becomes too slow and therefore rate determining. Teng et al. (2000) [45] have studied the growth mechanisms and rates of calcite as a function of supersaturation in situ by Atomic Force Microscopy (AFM). At $SI = 0.2$ they find spiral growth and growth at available molecular steps as the dominant growth mechanisms. At $SI = 0.4$ spiral growth and surface nucleation coexist. At $SI = 0.7$ crystal growth is clearly dominated by surface nucleation.

2.4. Sorption of trace elements at the calcite surface

Several types of reactions are described by the term "sorption of ions at a mineral surface". Ions can loose part of their hydration sphere and form covalent or ionic bonds with the

2. State of knowledge

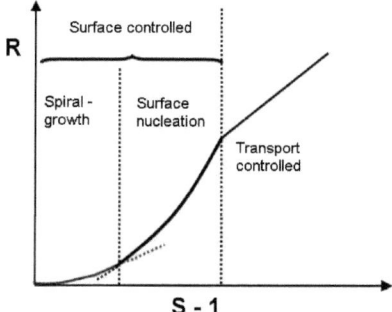

Fig. 2.6: Growth rate determining processes depending on supersaturation. (after [69])

surface. This type of adsorption is called chemisorption or inner–sphere adsorption. Ions can as well keep their hydration sphere and bond to the surface electrostatically or by hydrogen bonds. This type of adsorption is called physisorption or outer–sphere adsorption. Another type of sorption reaction frequently occurs in batch type adsorption experiments. The ions somehow become incorporated into the surface layers of the calcite crystal. This could happen, e.g., by ion exchange at the topmost crystal monolayer, by incorporation reactions that involve even deeper atomic layers of the crystal, or by subsequent dissolution reprecipitation reactions.

The sorption of trace elements especially metal cations onto calcite has been recognized by many researchers to be of major relevance for their mobility in soils and sediments. Correspondingly, a large amount of literature is available on this topic. Nevertheless, there is not yet agreement on the mechanisms that occur during sorption reactions. In many cases comparison of sorption and desorption experiments show that sorption reactions are partly irreversible. Studies of sorption kinetics often show that two reaction steps govern the sorption process. A fast initial reaction, which is usually interpreted as surface adsorption, and a second slow reaction, which is usually interpreted as an incorporation reaction. Such incorporation reactions are assumed to cause the partial irreversibility of sorption (e.g. [70, 71, 7]). The mechanism of such incorporation reactions is widely discussed. Suggestions range from ion exchange and subsequent solid state diffusion [72, 73, 71] to dissolution and reprecipitation reactions driven by the differing solubility of pure calcite

2. State of knowledge

and the solid solution resulting from the incorporation reaction (e.g., [74, 75]).

Early $^{45}Ca^{2+}$ calcite ion exchange experiments showed that more than one monolayer of calcite might be involved in the ion exchange reactions [71]. Stipp et al. (1998) [73] report rather unspecific surface mixing processes at dynamic equilibrium that range up to 10 atomic layers into the crystal. However, in situ AFM experiments demonstrated that no solid state diffusion is necessary to explain such phenomena. In the presence of solutions containing foreign ions calcite dissolves and on its place less soluble solid solutions reprecipitate [74, 75]. This could be shown to be the case in presence of Ba^{2+}, Sr^{2+}, Mn^{2+}, Cd^{2+}, and Mg^{2+}.

In principle ion diffusion through the bulk calcite structure can be observed at high temperatures [76]. Extrapolation of high temperature diffusivities to room temperature shows however that diffusion should be negligible at standard conditions. However, it cannot be completely excluded that the outermost few monolayers of a crystal behave different in that respect than the bulk crystal. For example, for Pb^{2+} adsorption at calcite sorption kinetics and desorption experiments indicate that incorporation may take place at calcite equilibrium. Linear combination fitting of x-ray absorption near edge structure (XANES) spectra has shown that after batch adsorption experiments at pH 7.3 and pH 9.4 up to 25 % of adsorbed lead likely belongs to an incorporated Pb^{2+} species [70]. At pH 8.3 no indication for Pb^{2+} incorporation into calcite has been found after a 2.5 years reaction period. Extended x-ray absorption fine structure (EXAFS) data indicates an inner–sphere adsorption species [77]. For trivalent lanthanides and actinides it has been observed that in the presence of Nd^{3+} calcite dissolves and a Nd-Ca-carbonate phase precipitates [78]. Adsorption experiments with Eu^{3+} and Cm^{3+} show that over longer timescales a completely dehydrated species can be observed by time resolved laser fluorescence spectroscopy (TRLFS), which is interpreted as an incorporation species [79, 19].

Geipel et al. (1997) [80] have studied uranyl adsorption at calcite. EXAFS data shows an adsorption complex very similar to the aqueous uranyl triscarbonato species. Therefore they conclude that uranyl does not adsorb at calcite as an inner–sphere complex. In a similar study Elzinga et al. (2004) [21] observed an uranyl triscarbonato species with a split equatorial coordination environment, which they interpret as a sorbed uranyl triscarbonato complex. They speculate that the splitting of the equatorial oxygen shell might be due to inner–sphere adsorption of the complex at the surface. Polarization dependent grazing incidence EXAFS measurements indicate that the orientation of the linear uranyl

2. State of knowledge

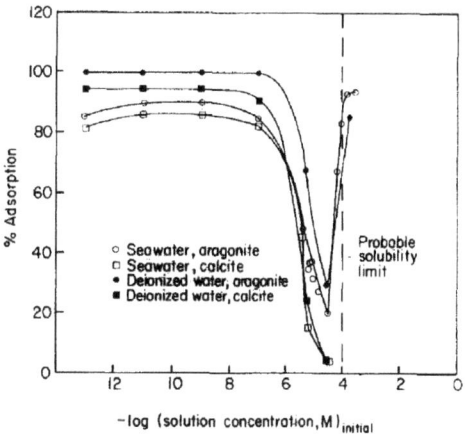

Fig. 2.7: Neptunyl(V) adsorption at calcite and aragonite in equilibrium with deionized and seawater [29].

ion adsorbed at the calcite surface is perpendicular or at a constant small angle tilt to the surface. This would agree well with adsorption at step edge [81].

Keeney-Kennicutt and Morse (1984) [29] have been the first to study neptunyl(V) adsorption at mineral surfaces over a large range of concentrations (10^{-13} M – 10^{-5} M). They report high affinity of neptunyl(V) for carbonate minerals. Adsorption decreases in the series: aragonite $\geq$ calcite > goethite $\gg$ MnO$_2$ $\approx$ clays. Some of their results on calcite and aragonite are shown in Figure 2.7. At equilibrium with atmospheric CO$_2$, after six hours reaction time, at low neptunyl concentrations, about 90 % of the neptunyl is adsorbed at the surface. Above a concentration of about 10^{-7} M adsorption decreases and reaches a minimum of about 5 % at a concentration of about 10^{-5} M. At concentrations of about 10^{-4} M the solubility limit is reached.

Zavarin et al. (2005) have studied the pH dependence of neptunyl(V) adsorption to calcite at an initial neptunyl(V) concentration of 10^{-7} M [28] in equilibrium with calcite and atmospheric CO$_2$. They observe maximum adsorption at pH 8.5, decreasing towards lower and higher pH. They model their experimental data, shown in Figure 2.8, based on the constant capacitance SCM by Pokrovsky and Schott [62] using two surface species: >CaCO$_3$NpO$_2^0$ and >CaCO$_3$NpO$_2$CO$_3^{2-}$.

2. State of knowledge

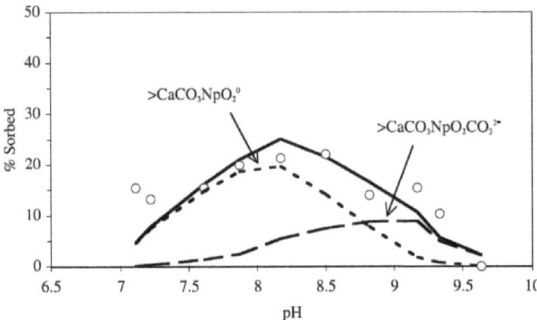

Fig. 2.8: PH dependence of neptunyl(V) adsorption at calcite [28]. (Experimental Data – circles; and model curves: individual surface species – dashed lines, sum – solid line)

It is very difficult to prove experimentally if incorporation indicated in batch type sorption experiments is due to thermodynamically driven dissolution and reprecipitation reactions, or ion exchange and diffusion, or if it is due to any other unwanted perturbation of the calcite equilibrium, e.g., during the addition of the sorbent, by temperature or CO_2 partial pressure changes, or by evaporation of solution over longer timescales. In case of unwanted equilibrium perturbations, incorporation could rather be an experimental artifact than a reaction relevant in natural systems. Therefore, incorporation is often studied specifically by means of coprecipitation experiments. The aim of such experiments is to investigate if structural incorporation is possible and if incorporation is thermodynamically favorable. Synthesis of higher amounts of incorporation species allows to study their structure in more detail.

2.5. Coprecipitation of trace elements with calcite

As mentioned in the introduction (1.1) there are numerous studies on coprecipitation of trace elements with calcite (see [5, 6] for summaries).
Concerning the actinides, uranium uptake is most investigated. Uranium(IV) in a natural calcite sample has been studied by Sturchio et al.(1998) [20]. They use EXAFS spectroscopy to investigate the local structural environment of uranium(IV) in calcite and find that uranium(IV) occupies a stable position in the calcite structure. They have not been

2. State of knowledge

able to identify the exact mechanism providing charge compensation upon the substitution of Ca^{2+} for U^{4+}. Reeder and co–workers report on the incorporation of hexavalent uranyl into artificially synthesized calcite [22, 23, 24]. Their results reflect that the local environment of uranyl in calcite is not compatible with the calcite structure. They find intrasectoral zoning patterns that show that uranyl is incorporated into calcite preferentially at acute steps on the calcite (104)–face. This reflects the importance of the surface morphology for trace element incorporation and indicates non–equilibrium conditions during coprecipitation. Kelly et al. studied the structural environment of uranyl in natural calcite [25, 26]. In the natural sample the local environment of uranyl in calcite is compatible with the calcite structure, showing a fourfold monodentate carbonate coordination in the equatorial plane of the linear uranyl ion, similar to calcium in calcite. They propose that the axial uranyl oxygen atoms substitute two carbonate groups. They cannot tell from their data how the charge excess caused by this substitution is balanced.

It has been shown that the trivalent lanthanide Eu^{3+} and the trivalent actinides Cm^{3+} and Am^{3+} can be incorporated into the calcite crystal structure [19, 18, 82]. Comparative Eu^{3+} coprecipitation experiments in presence and absence of sodium (replacing the background electrolyte NaCl with KCl) indicate that charge compensation is provided by the coupled substitution mechanism: $2\ Ca^{2+} \leftrightarrow Eu^{3+}+Na^+$ [82]. However, thermodynamic modeling studies report other substitution mechanisms for Eu^{3+} incorporation into calcite to be the more likely [17].

To compare experimental results it is often sufficient to describe the trace cation (M^{x+}) uptake in terms of the homogeneous Henderson-Kracek partition coefficient, D. For calcite this is:

$$D = \frac{X_M/X_{Ca}}{[M^{x+}][Ca^{2+}]} \qquad (2.4)$$

where X_M and X_{Ca} are the molar fractions of the trace element ion and calcium in the precipitating solid and $[M^{x+}]$ and $[Ca^{2+}]$ are the corresponding total concentrations in the equilibrium or steady state solution. Solution composition needs to be constant and the precipitating solid homogeneous in order to calculate this partition coefficient.

In the simplest case Ca^{2+} ions are substituted by other divalent cations M^{2+} like Mg^{2+}, Fe^{2+}, Zn^{2+}, Mn^{2+}, Sr^{2+}, Ba^{2+}, or Cd^{2+} that form carbonates with the same chemical formula as calcite, MCO_3. The formation of binary solid solutions of this kind has been intensively studied. In these cases the empirical Henderson-Kracek partition coefficient is equal to a thermodynamic partition coefficient.

2. State of knowledge

In case of substitution by ions of different charge or complex ions like uranyl or neptunyl, coupled substitution mechanisms have to be taken into account that provide charge balance. The examples of incorporation studies mentioned above demonstrate that it often is very difficult or impossible to elucidate the substitution mechanism experimentally. The pure trace element phase that is one end member of the mixing series is often unknown. In these cases empirical partition coefficients indicate if the trace metal ion has a higher affinity for the solid phase or for the liquid phase. Such partition coefficients are, however, not thermodynamically meaningful. Other expressions need to be considered to handle more complex substitution mechanisms. How they can be derived is shown in the next section.

2.5.1. Solid solution – aqueous solution equilibria

A solid solution is a single mineral phase which exists over a wide range of chemical compositions. Almost all minerals are able to tolerate variations in their chemistry. The chemical variation greatly affects the stability and behavior of the mineral. Thermodynamic stability of a solid solution is best described by the excess Gibbs free energy of mixing, ΔG_{mix}. ΔG_{mix} describes the difference between the actual formation energy of the solid solution, ΔG, from the hypothetical formation energy of a physical mixture of the two endmember phases, ΔG_{ideal}:

$$\Delta G_{mix} = \Delta G - \Delta G_{ideal} = \Delta H_{mix} - T\Delta S_{mix}. \tag{2.5}$$

As the incorporation of foreign ions into a crystal structure usually causes strain additional energy is needed to accomplish the incorporation: the enthalpy of mixing, ΔH_{mix} is usually positive. On the other hand incorporation of foreign ions into a crystal structure causes disorder. Therefore the entropy of mixing, ΔS_{mix}, is also positive. This makes the stability of solid solutions highly dependent on the temperature, T. While high temperatures stabilize solid solutions over the whole series of possible compositions at room temperature often only trace amounts of a foreign ion can be incorporated into a host crystal structure. Equations for the description of solid solution – aqueous solution equilibria shown here have been collected from Tesoriero and Pankow (1996) [83], Glynn (2000) [84], and Shtukenberg et al. (2006) [85].

For the description of thermodynamic equilibrium between a solid solution of the arbitrary phases A and B and an aqueous solution, a thermodynamic partition partition coefficient

2. State of knowledge

can be formulated:
$$D = \frac{X_A/X_B}{ICP_A/ICP_B} \qquad (2.6)$$

where X_A and X_B are the mole fractions of the two endmember phases A and B in the solid solution ; $X_B = 1 - X_A$. ICP_A and ICP_B are the ion concentration products of the two endmembers in the aqueous solution. They are related to the ion activity products IAP_A and IAP_B via the product of the activity coefficient in aqueous solution γ_A and γ_B.

$$IAP_A = ICP_A \cdot \gamma_A \qquad (2.7)$$
$$IAP_B = ICP_B \cdot \gamma_B \qquad (2.8)$$

With the solubility products K_A and K_B of the two endmembers A and B, law of mass action equations can be defined:

$$IAP_A = K_A X_A f_A \qquad (2.9)$$
$$IAP_B = K_B X_B f_B \qquad (2.10)$$

where the solubility products and the mole fractions of the endmembers are linked via solid solution activity coefficients f_A and f_B to the corresponding ion activity products. Introducing these equations into equation 2.6 the partition coefficient D becomes:

$$D = \frac{K_B f_B \gamma_A}{K_A f_A \gamma_B} \qquad (2.11)$$

Guggenheim (1937) [86] has described the dependence of the excess Gibbs free energy of mixing, ΔG_{mix}, on the composition of a chemical mixture by an expansion series:

$$\Delta G_{mix} = X_A X_B RT(a_0 + a_1(X_A - X_B) + a_2(X_A - X_B)^2 + \ldots) \qquad (2.12)$$

where R and T are the gas constant and the absolute temperature and a_i are fitting parameters, the so-called Guggenheim parameters. This expression can be used to describe the dependency of the solid solution activity coefficients on the solid solution composition.

$$\ln f_B = X_A^2[a_0 - a_1(3X_B - X_A) + a_2(X_B - X_A)(5X_B - X_A) + \ldots] \qquad (2.13)$$
$$\ln f_A = X_B^2[a_0 + a_1(3X_A - X_B) + a_2(X_A - X_B)(5X_A - X_B) + \ldots] \qquad (2.14)$$

In many cases it is sufficient to use one or two Guggenheim parameters in order to reproduce the solid solution activity coefficients. Solid solutions described by one Guggenheim parameter are called regular solid solutions. For regular solid solutions the activity

2. State of knowledge

coefficients behave symmetrically across the mixing series. Using a second Guggenheim parameter, an asymmetry in the dependence of the solid solution activity coefficients on the mole fractions is introduced. Solid solutions that can be described by two Guggenheim parameters are called subregular. For a regular solid solution the solid solution activity coefficients in equation 2.11 can be replaced and the partition coefficient becomes:

$$D = \frac{K_B \gamma_A}{K_A \gamma_B} \exp(a_0(X_B^2 - X_A^2)). \tag{2.15}$$

A phase diagram that relates the equilibrium composition of a solid solution to the composition of the corresponding aqueous solution has been developed by Lippmann (1980) [87]. The total solubility product, $\Sigma\Pi$, for a certain solid composition is given by the "solidus" equation:

$$\Sigma\Pi = K_A X_A f_A + K_B X_B f_B. \tag{2.16}$$

The total solubility product, $\Sigma\Pi$, for the corresponding solution composition is given by the "solutus" equation:

$$\Sigma\Pi = 1 / \left(\frac{X_{A(aq)}}{K_A f_A} + \frac{X_{B(aq)}}{K_B f_B} \right) \tag{2.17}$$

where $X_{A(aq)}$ and $X_{B(aq)}$ are defined as:

$$X_{A(aq)} = IAP_A / (IAP_A + IAP_B) \quad \text{and} \tag{2.18}$$

$$X_{B(aq)} = IAP_B / (IAP_A + IAP_B); \tag{2.19}$$

$$X_{B(aq)} = 1 - X_{A(aq)} \tag{2.20}$$

In the Lippmann diagram two graphs are plotted. The solidus graph: $\Sigma\Pi \mapsto X_B$ and the solutus graph: $\Sigma\Pi \mapsto X_{B(aq)}$. Solid solutions and aqueous solutions in equilibrium with each other have the same $\Sigma\Pi$ value on the ordinate. The corresponding equilibrium solid solution and aqueous solution composition can be read from the abscissa. An example for a Lippmann diagram is shown in Figure 2.9.

The thermodynamic expressions to describe solid solution – aqueous solution equilibria are well established. The challenge when studying solid solution formation from aqueous solution is to find out to which extent they can be used. That means to find out to which extent experiments reflect equilibrium conditions. It is common practice to perform experiments at low supersaturation and low growth rates and to interpret the results by equilibrium thermodynamics although it is known that distribution coefficients depend

2. State of knowledge

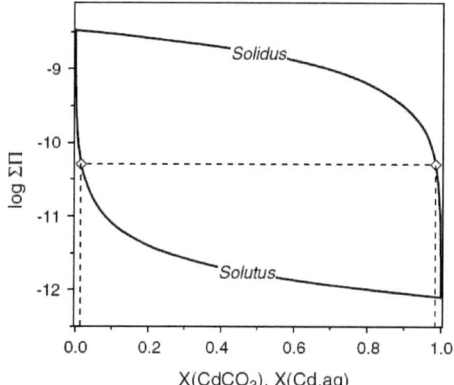

Fig. 2.9: Lippmann diagram for the calcite otavite (CdCO$_3$) solid solution [74]. Solids and solutions in equilibrium with each other are linked by horizontal lines. The corresponding solid and solution composition can be read from the abscissa.

on supersaturation. With increasing supersaturation the solid composition becomes more and more similar to the solution composition. For solid solutions involving only simple substitution mechanisms the distribution coefficient approaches one. Lowering the supersaturation the distribution coefficient approaches its equilibrium value.

It is well known that kinetic effects influence the coprecipitation of trace elements with calcite. Examples are, the dependency of the partition coefficients on supersaturation or growth rate and sectoral or intrasectoral zoning phenomena that have been reported [22, 42, 47] and that cannot be described by thermodynamics.

2.5.2. Influence of trace element incorporation on crystal growth

Trace metal ions are often adsorbed selectively onto different crystal faces and retard their growth rate [35]. It is not necessary for the cations to achieve total face coverage to cause retardation. As seen in Figure 2.10 three sites may be considered at which trace element cations can adsorb and disrupt the flow of growth layers across the crystal faces. Trace cations adsorbing at kink or step sites can retard the growth rate even at very low concentrations (Figure 2.10 (a) and (b)). Cations adsorbing to the plane crystal face need to

2. State of knowledge

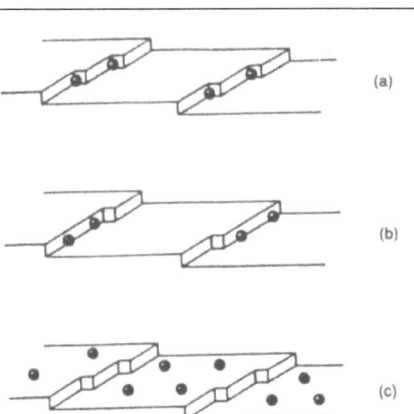

Fig. 2.10: Sites for trace element-adsorption on growing flat crystal faces [35].

be present at much higher concentrations to have any influence (Figure 2.10 (c)). Such impurities influence the growth rate only when the face coverage becomes so dense that the formation of surface nuclei is hindered.

If the influence of trace element incorporation on growth rate is face selective, coprecipitation can change the crystal habit [35].

Another approach to describe the effect of incorporation on growth rate is considering the energy barrier that needs to be overcome during solid solution growth on a substrate that is due to the strain caused by lattice misfit between the substrate and the growing solid solution [85].

Many studies are reported in literature that characterize the calcite surface. Previous calcite SCMs describe the calcite surface charging behavior including inner–sphere complexes at the calcite surface [62, 58], ignoring the fact that it has been observed that calcium most likely forms outer–sphere at the calcite surface [88] and that x–ray reflectivity mea-

2. State of knowledge

surements show that there are most likely no (< 9 %) inner–sphere complexes on terrace planes on the calcite (104)–face [48]. One aim of this study is to verify the results of the x–ray reflectivity study under more extreme solution conditions and investigate the full 3D structure of the calcite–(104)–water interface by means of surface diffraction. The results of this study should then serve as structural constraints for a new SCM for calcite based on zetapotentials measured under known, well controlled conditions.

Previous studies report the concentration– and pH dependence of neptunyl(V) adsorption at the calcite surface [29, 28], but a systematic investigation of both parameters together is still lacking. Aim of the adsorption study presented here is to close this gap, to provide spectroscopic insight into the structure of the adsorption complexes, and the sorption mechanism.

The coprecipitation of uranyl(VI) has been thoroughly investigated [23, 24, 22]. No comparable studies about coprecipitation of neptunyl(V) with calcite have been available before this study. Therefore a coprecipitation study has been performed to quantify the neptunyl incorporation into calcite and characterize the incorporation species' structure. Additional information about the calcite precipitation rates obtained from coprecipitation experiments can be used to learn about the surface processes involved during coprecipitation.

Chapter 3

Experimental details

3.1. Analytical methods

3.1.1. Scanning electron microscopy (SEM)

In order to obtain images of small calcite crystallites, e.g., crystals used for zetapotential measurements, crystal seeds for mixed flow reactor (MFR) experiments, or the reaction products of MFR experiments, SEM images have been taken on a CamScan CS44FE SEM in secondary electron mode at an acceleration voltage of 19–20 kV. If necessary crystals are coated with chromium to avoid charging of the non–conducting calcite in the electron beam.

3.1.2. X–ray photoelectron spectroscopy (XPS)

XPS is a highly sensitive method to study the surface chemistry of minerals. The energy of photoelectrons excited from the surface atoms by Al or Mg K_α radiation carries information about the chemical state of the surface atoms. XPS is performed in ultra high vacuum on a PHI 5600ci instrument.

3.1.3. Determination of the specific surface area by N_2–BET

The specific surface area of commercial calcite powder (Merck calcium carbonate suprapur) used in adsorption experiments and as crystal seeds in MFR experiments has been measured by N_2–BET. BET measurements are performed in a Quantachrome Autosorb

3. Experimental details

Automated Gas Sorption System. Specific surface area of the calcite powder used in MFR and adsorption experiments is 1.28 m^2/g. Standard deviation of BET measurements has been estimated to be 8.6 %.

3.1.4. X-ray powder diffraction (powder XRD)

Powder XRD has mainly been used to verify if crystal powders used for the experiments are pure calcite or contain other mineral phases. Powder XRD measurements are performed on a Bruker D 5000 diffractometer. Merck calcium carbonate suprapur shows typical calcite powder patterns, before MFR and Adsorption experiments as well as after MFR experiments. Therefore we conclude that Merck calcium carbonate suprapur consists only of calcite (100 ±1%).

3.1.5. Inductively coupled plasma mass spectrometry (ICP–MS)

Neptunium and calcium solution concentrations of samples from all MFR experiments involving radioactive material are measured using an ICP–MS ELAN 6100 inductively coupled plasma mass spectrometer. From measurements of NIST standards the standard deviation of ICP–MS results has been estimated to be 6.6 %.

3.1.6. Inductively coupled plasma optical emission spectrometry (ICP–OES)

Calcium solution concentrations of samples from inactive MFR experiments are measured on a Perkin Elmer Optima 2000 DV inductively coupled plasma optical emission spectrometer. From measurements of NIST standards the standard deviation of ICP–OES results has been estimated to be 4.1 %.

3.1.7. Liquid scintillation counting (LSC)

Neptunium-237 ($t_{1/2} = 2.144 \cdot 10^6$ a) solution concentrations of samples from adsorption experiments are measured by liquid scintillation counting (LSC) on a Perkin Elmer Tricarb liquid scintillation counter in the Packard Ultima Gold XR scintillation cocktail. Alpha/Beta discrimination is used to differentiate between neptunium–237 (α–radiation) and the daughter nuclide protactinium–233 (β–radiation). The spectra are analyzed using

3. Experimental details

the Packard SectraWorks 1.0 software. Alpha counts are summed over an energy range between 120 and 300 keV.

3.1.8. Error propagation calculations

If not otherwise specified errors for parameters are calculated by error propagation calculations according to the following relation: The error, ΔF, of a quantity, F, defined as a function of n variables, $F = F(x_1, x_2, \ldots, x_n)$ is related to the errors of the variables Δx_1, $\Delta x_2, \ldots, \Delta x_n$ by:

$$\Delta F = \sum_{i=1}^{n} \left| \frac{\partial F}{\partial x_i} \right| \Delta x_i . \tag{3.1}$$

3.2. X-ray absorption fine structure (XAFS)

As x–ray absorption fine structure (XAFS) measurements play a central role in the work presented here, the basic principles of this methods will be briefly explained. Details about measurements and data analyses will be given individually in the following sections.

XAFS refers to the modulation of the probability of an atom to absorb x–rays, depending on x–ray energy. This probability depends on the the type, the oxidation state and the structural environment of an atom. Therefore XAFS spectra can be used to find out the number, type, and distance of neighboring atoms, and the valence, the type of bonds, and the occupation of energy levels of the examined atom type in a distinct chemical environment. As XAFS is only sensitive to the nearest environment of an atom, only short range order is needed in the sample. XAFS studies can be performed on trace levels down to abundances of only several hundred ppm of the examined element in the sample.

In XAFS measurements monochromatic X-rays originating from a synchrotron light source are directed onto the sample (see Figure 3.1). The x–ray energy is selected and varied by adjusting the Bragg-angle of a double crystal monochromator. When the x–ray energy (E) exceeds the binding energy of a core level electron (E_0) of the examined atom type it promotes, according to the photoelectric effect, the electron into the continuum and an absorption edge is observed. The measured absorption around and above the absorption edge relative to the x–ray energy gives the XAFS spectrum. Illustrated in Figure 3.1 are two different ways to measure such spectra. In the transition mode the intensity of the x–ray beam is measured before (I_0) and after (I) the sample. X-rays are absorbed according

3. Experimental details

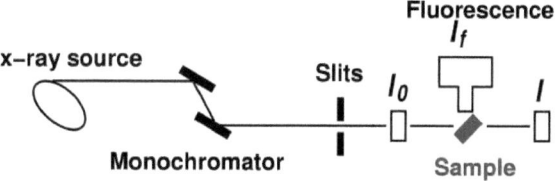

Fig. 3.1: Experimental setup for XAFS measurements [89]

to the Lambert-Beer law:

$$I = I_0 \, e^{-\mu t} \qquad (3.2)$$

with, t, the thickness of the sample and, μ, the absorption coefficient, and therefore the absorption can be calculated as:

$$\mu(E) \cdot t = \ln \frac{I_0}{I} \qquad (3.3)$$

By the absorption of photons core-level electrons are promoted to the continuum. When a higher energy electron relaxes to fill the resulting core hole an fluorescent x–ray is emitted. This leads to another way to measure the absorption. Because fluorescence is directly proportional to absorption measuring I_0 and the Intensity of the resulting x–ray fluorescence (I_f) allows calculating the absorption:

$$\mu(E) \propto \frac{I_f}{I_0} \qquad (3.4)$$

This method is called fluorescence mode.

Figure 3.2 shows what happens to the promoted electron above the absorption edge, using a carbon 1s (K) edge spectrum as an example. For x–ray energies close to E_0 the electron can be promoted into unoccupied bond states of the excited atom (Region A in Figure 3.2). This is the reason for a peak in absorption the so–called whiteline. The energy position of E_0 or the whiteline give information about the valence state of the excited atom. For higher energies the electron propagates with the kinetic energy $E_{kin} = E - E_0$ into the continuum. In order to understand the modulations in the absorption above the edge it is necessary to imagine the photoelectron as a spherical electron wave that is scattered back from neighboring atoms. Region B in Figure 3.2 marks the energy range in which the electron wave is scattered on a number of neighbor atoms before returning to the exited atom, the so called multiple scattering region. Regions A and B together mark

3. Experimental details

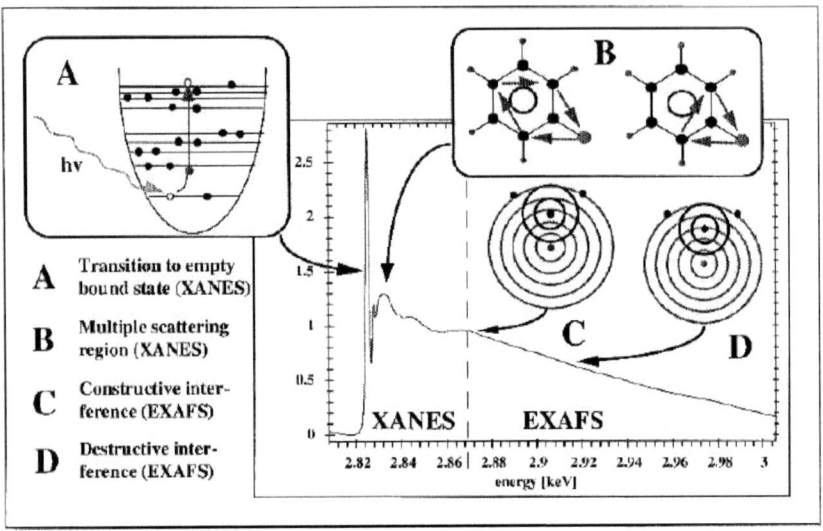

Fig. 3.2: The effects that lead to the oscillations of an XAFS spectrum above the absorption edge.

the XANES or x–ray Absorption Near Edge Structure region. It is difficult to predict or model the absorption behavior in this region. Analysis of XANES spectra is therefore often done by comparison of measured spectra with reference spectra, a method called fingerprinting. At about the energy where the wavelength of the photoelectron becomes shorter than the distance between the excited atom and its nearest neighbor the EXAFS region begins. EXAFS stands for Extended X–ray Absorption Fine Structure. When there is constructive interference between the electron wave going out from the excited atom and the wave scattered back from an neighboring atom, the absorption probability is higher (Region C in fig. 3.2) than when they interfere destructively (Region D in fig. 3.2). The interference between the outgoing and the backscattered wave leads to the oscillating EXAFS pattern which can be modeled. By fitting a theoretical model to the measured spectrum information about type and number of neighboring atoms and the inter atomic distances can be derived.

Before the fitting procedure the spectrum has to be extracted from the raw data. To

3. Experimental details

separate the EXAFS first the background absorption of all the other elements in the sample matrix is subtracted. Then a smooth background function ($\mu_0(E)$) is subtracted and the difference is divided by the edge step ($\mu_0(E_0)$) to normalize the resulting $\chi(E)$ to one absorption event.

$$\chi(E) = \frac{\mu(E) - \mu_0(E)}{\mu_0(E_0)} \quad (3.5)$$

The resulting function is the EXAFS. It is usually not displayed relative to the x–ray energy, E (eV), but relative to the electron wave number, k (Å^{-1}). The conversion from E to k, so called k–space, is done according to the equation:

$$k = \sqrt{\frac{2m(E - E_0)}{\hbar^2}} = \sqrt{\frac{2m \cdot E_{kin}}{\hbar^2}} \quad (3.6)$$

where m is the resting mass of an electron. $\chi(k)$ is often weighted by k, k^2, or k^3 to amplify the intensity of the oscillations at high k especially those from the lighter atoms. For spectra measured at the K–edge of an element $\chi(k)$ can be modeled after this EXAFS equation:

$$\chi(k) = S_0^2 \sum_j \frac{N_j f_j(k) e^{-2k^2 \sigma_j^2} e^{-2R_j/\lambda(k)}}{kR_j^2} \sin(kR_j + \delta_j(k)) \quad (3.7)$$

$f_j(k)$ and $\delta_j(k)$ are backscattering amplitude and phase shift functions, which describe the backscattering properties of the neighboring atoms. From these functions the type of neighboring atoms (atomic number Z) can be determined. $\lambda(k)$ is the mean free path of the photoelectron. It is typically between 5 and 30 Å and depends on k. The sum is taken over j "shells" or scattering paths. A shell consists of similar neighboring atoms at the same distance from the central atom. N_j is the number of atoms in shell j, and R_j is the distance of the atoms in the shell j from the central atom. σ_j^2 is the EXAFS Debye-Waller factor or mean square displacement of atoms in shell j and it describes both structural disorder and thermal vibrations. S_0^2 is the amplitude reduction factor; it depends only on the central atom and is equal for all the different shells.

The Fourier transform from k–space to distance or R–space gives a modified pair distribution function of the central absorbing atom. Fits in this work are performed in R–space. The results from least square fits of a calculated coordination structure model spectrum to the measured spectrum provide N, R, Z, and σ^2 for each shell. From these values conclusions about the structural environment of the examined element in the sample can be drawn. [89]

3.3. pH measurements

A high precision is desired for the pH measurements during adsorption experiments and during the calcite equilibration procedure preceding zetapotential measurements and surface diffraction, as pH is the parameter used to specify if equilibrium is achieved. Measurements are performed using a Thermo Fisher Scientific ORION ROSS semi–micro combination pH electrode and the Orion pH–Meter model 920 A. The pH electrode has frequently been calibrated using five Merck Titrisol pH buffer solutions in a pH range from 6 to 10. Using this setup the standard deviation of pH measurements is reduced to 0.08–0.12 pH units.

Various pH–meters and electrodes have been used for pH measurements during all other experiments.

3.4. Calcite equilibration method

For many of the experiments performed during this study it is important to ensure equilibrium between calcite, solution, and a surrounding gas phase. Equilibration of calcite suspensions has been achieved in an experimental setup like the one shown in Figure 3.3 or slight modifications of it. Gas from a reservoir is wetted in a washing flask and then distributed to the reaction vessels that contained calcite suspensions. During the equilibration procedure suspensions are permanently percolated with gas. This is done in a way that ensures high contact areas between gas and liquid phase and keeps the calcite suspended to make equilibration as quick as possible. In the neptunyl adsorption experiments the gas has been collected after the reaction vessels in a final bottle as shown in Figure 3.3 to avoid contamination. This has not been necessary in the other experiments. In experiments with air the gas bottle has been replaced by an electric air pump. The setup has been modified for some experiments with respect to numbers and sizes of reaction vessels. Depending on the desired solution condition gases with CO_2 partial pressure ranging from $10^{-5.2}$ (6 ppm) to one (pure CO_2) are used at atmospheric pressure. Solutions are composed of MilliQ water (18.2 MΩ), HCl, or NaOH, and NaCl. With these gas phases and solution compositions a pH range from 5.8 to 10.3 can be covered at an ionic strength < 0.15 M. Equilibration has been run until the equilibrium pH calculated by thermodynamic modeling is reached. Depending on solution and gas composition the time required for equilibration varied from

3. Experimental details

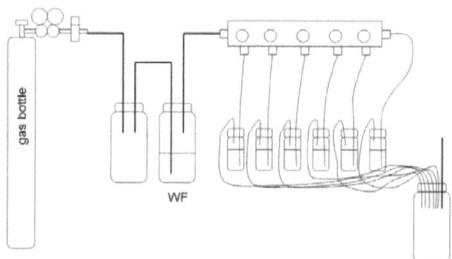

Fig. 3.3: The experimental setup for the equilibration of calcite suspensions with a gas phase and for adsorption experiments. Gas from a gas bottle is wetted in a washing flask (WF) and then distributed to the six reaction vessels. In the adsorption experiments the gas is collected in a final bottle to avoid contamination. In experiments with air the gas bottle is replaced by an electric air pump.

one day to about one month.

3.5. Zetapotential measurements

Zetapotential measurements described in this section are those which resulted in the datasets chosen for subsequent surface complexation modeling. Numerous preliminary zetapotential measurements have been performed to examine how to obtain reasonable results with the two different methods. Main experimental limitations of these methods are set by ionic strength and particle size in PALS measurements and by ionic strength and carbonate concentration in the streaming potential measurements.

3.5.1. Phase analyses light scattering (PALS)

PALS measurements are performed on a Brookhaven Instruments PALS zetapotential analyzer. Measurements are performed on Merck calcium carbonate p a ground in a ball mill to particle sizes < 1 μm. Powder XRD showed that Merck calcium carbonate p.a.

3. Experimental details

consists only of calcite (± 1 %). The grinding is necessary to assure that the particles stay suspended during the measurements and do not sediment. Sedimentation of the particles during the measurements adulterates PALS results. Equilibrium solutions have been prepared by the method described above (cp. section 3.4). Equilibrium pH of solutions in equilibrium with pure CO_2 ranges from 5.8 to 6.8. This has been achieved by varying the initial acid or base concentration from 0.06 M HCl to 0.1 M NaOH. NaCl is added to maintain the equilibrium ionic strength between 0.10 and 0.11 M. In equilibrium with air (360 ppm CO_2) pH ranged from 7.5 to 9.7. Solution composition is varied from 0.1 M HCl to 0.1 M NaOH. Ionic strength ranges from 0.10 to 0.15 M. In equilibrium with N_2 (6 ppm CO_2) pH ranges from 8.3 to 10.3. Solution composition is varied from 0.07 M HCl to 0.07 M NaOH. Ionic strength ranges from 0.10 to 0.11 M. Zetapotentials are calculated from the electrophoretic mobility according to the Smoluchowski equation:

$$u = \frac{\epsilon \epsilon_0 \zeta}{\eta} \tag{3.8}$$

where u is the electrophoretic mobility ($m^2/(Vs)$), ϵ is the permittivity of water (78.54), ϵ_0 is the permittivity of free space ($8.854 \cdot 10^{-12}$ F/m), ζ is the Zeta potential (in V), and η is the dynamic viscosity of water (0.00089 Pa · s). The Smoluchowsky equation is derived under the assumption that $\kappa a \gg 1$, with κ being the reciprocal Debye length (m^{-1}), and a the particle diameter (m). In the system considered in this study this value is about 10^2, so the equation should be well applicable. κ depends on the ionic strength, I (mol/L), and can be calculated by:

$$\kappa = \sqrt{\frac{2000 N_A e^2 I}{\epsilon \epsilon_0 k_B T}}. \tag{3.9}$$

N_A is the Avogadro constant ($6.02214 \cdot 10^{23}$ mol^{-1}), e is the elementary charge of an electron ($1.602177 \cdot 10^{-19}$ C), k_B is the Boltzmann constant ($1.38066 \cdot 10^{-23}$ J/K), and T is the absolute Temperature (in K).

In the PALS zetapotential analyzer the electrophoretic mobility is measured from the phase shift between laser light scattered from particles moving between two electrodes by an angle of 20° and unscattered laser light. Standard deviation of electrophoretic zetapotential measurements is about 1.2 mV. It increases for zetapotential close to zero and decreases with increasing norm of the zetapotential. It must be assumed however that the systematic error is much bigger, ~ 5 mV.

3. Experimental details

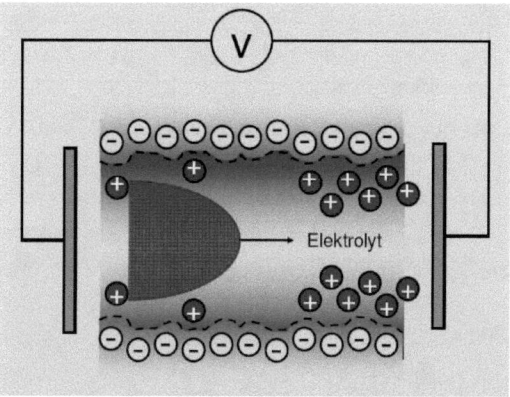

Fig. 3.4: Principle of streaming potential measurements in the SurPASS elctrokinetic analyzer.

3.5.2. Streaming potential measurements

Streaming potential measurements are performed on an Anton Paar SurPASS electrokinetic analyzer. Coarse crystallites > 25 μm in diameter are ground from natural calcite var. Iceland spar from Mexico Chihuahua from Ward's natural science. XPS shows that the natural calcite is very clean and shows no measurable contaminations with other metals or anions. Measurements are performed in the SurPASS powder sample cell where the calcite powder is trapped between two filter plates. The principle of streaming potential measurements is shown in Figure 3.4. Due to a differential pressure, p, between the two ends of the sample cell, the electrolyte flows through the cell. The flow of electrolyte shears ions from the electric double layer at the mineral solution interface. The transport of these ions produces a potential difference, U (mV), between the two sides of the sample cell. This potential difference is measured as a function of pressure, p (mbar). The sample thickness and the pressure range used need to be adjusted to get a linear dependence between potential and pressure. Then the zetapotential can be calculated by:

$$\zeta = \frac{dU}{dp} \cdot \frac{\eta}{\epsilon \epsilon_0} \cdot ec \tag{3.10}$$

where ec (mS/m) is the electric conductivity of the electrolyte solution. Standard deviation of streaming potential measurements is about 1.5 mV. For this method it is also likely that the systematic error is much larger, $\sim$ 5 – 10 mV.

3. Experimental details

Non–equilibrium solutions are used for the streaming potential measurements. pH is varied between 5.5 and 11 by titration either with 0.1 M HCl starting from pH 11 or 10.7 or by titration with 0.1 M NaOH starting from pH 5.5. $CaCl_2$ concentration is varied from 0 to 5 mM. One titration has been performed with 5 mM Na_2CO_3. Ionic strength is adjusted to 0.01 M NaCl. Two datasets have been collected at IS = 0.001M and IS = 0.1 M. Measurements with elevated carbonate concentrations and correspondingly also solutions in equilibrium largely produced unreasonable results, probably due to the big pressure changes during the measurements that cause changes in CO_2 solubility. Therefore equilibrium solutions cannot be studied with this method; only one dataset with carbonate in solution seemed reliable. For this experimental setup the solution to surface ratio is very high (2L / ∼0.1 m^2); therefore the influence of calcite dissolution on the solution composition can be neglected. The calcium concentration in solution does not increase by more then 10^{-5} M during a whole titration procedure. However, in principle, the calcite must be dissolving during a titration. It cannot be excluded that the dissolving calcite surface shows a different charging behavior than a stable one, but above a pH of 5.5 calcite is known to dissolve slowly [60]. Therefore, this pH has been chosen as the lower limit of the experimental pH range.

3.6. Surface diffraction

3.6.1. Basic principles of surface diffraction

Surface diffraction is based on the principle that information about the quasi 2D surface or interface structure is contained in lines of scattering intensity in reciprocal space perpendicular to the crystal surface the so–called crystal truncation rods (CTRs) [90], while information about the 3D bulk structure of a crystal is gathered in points in reciprocal space that define the reciprocal lattice of the crystal.

To explain this it is necessary to take a look at how x–rays interact with matter. For many purposes it is convenient to describe scattering from a atom by the atom form factor, f^0, which is:

$$f^0(\mathbf{Q}) = \int \rho(\mathbf{r})e^{i\mathbf{Q}\cdot\mathbf{r}}d\mathbf{r}. \tag{3.11}$$

The integral is taken over all space, $\rho(\mathbf{r})$ is the electron density at point $\mathbf{r}$. $\mathbf{Q}$ is the scattering vector or momentum transfer. It is defined as the difference between the wave

3. Experimental details

vectors of the incident and outgoing, scattered x–rays, both described as plane waves with wave vectors $\mathbf{k_{in}}$ and $\mathbf{k_{out}}$: $\mathbf{Q} = \mathbf{k_{out}} - \mathbf{k_{in}}$. For elastic scattering, as we assume it to be in surface diffraction, $|\mathbf{k_{in}}| = |\mathbf{k_{out}}|$ and $|\mathbf{Q}| = 2|\mathbf{k_{in}}|\sin\theta$ for an angle of 2θ between the two wave vectors $\mathbf{k_{in}}$ and $\mathbf{k_{out}}$. A sketch of this geometry is shown in Figure 3.8.

The diffraction behavior of crystalline material results from the interference of x–rays scattered on electrons of many atoms in a periodic structure. The scattering length of one unit cell of this periodic structure is given by the unit cell structure factor, F_{uc}. It is defined as:

$$F_{uc}(\mathbf{Q}) = \sum_j f_j(\mathbf{Q}) \exp(2\pi i \mathbf{Q} \cdot \mathbf{r_j}) \exp\left(-\frac{B_j Q^2}{4}\right). \tag{3.12}$$

The sum is taken over j atoms with atom form factors f_j. The second exponential term in the sum in equation 3.12 contains B_j the Debye–Waller factors of the atoms. It describes smearing out of electron density due to thermal vibrations of the atoms. $\mathbf{r_j}$ are the positions of the atoms in the unit cell in crystal coordinates defined by the basis vectors $(\mathbf{b_1}\ \mathbf{b_2}\ \mathbf{b_3})$.

$$\mathbf{r_j} = (\mathbf{b_1}\ \mathbf{b_2}\ \mathbf{b_3}) \begin{pmatrix} x_j \\ y_j \\ z_j \end{pmatrix}. \tag{3.13}$$

In surface diffraction the basis vectors are often defined in a way that $\mathbf{b_1}$ and $\mathbf{b_2}$ lie in the surface, while $\mathbf{b_3}$ points out of the crystal. $\mathbf{Q}$ is given in reciprocal crystal coordinates.

$$\mathbf{Q} = (h\ k\ l) \begin{pmatrix} \mathbf{b_1^*} \\ \mathbf{b_2^*} \\ \mathbf{b_3^*} \end{pmatrix} \tag{3.14}$$

The reciprocal basis vectors are defined in a way that:

$$\begin{pmatrix} \mathbf{b_1^*} \\ \mathbf{b_2^*} \\ \mathbf{b_3^*} \end{pmatrix} \cdot (\mathbf{b_1}\ \mathbf{b_2}\ \mathbf{b_3}) = \begin{pmatrix} \mathbf{b_1^*}\mathbf{b_1} & \mathbf{b_1^*}\mathbf{b_2} & \mathbf{b_1^*}\mathbf{b_3} \\ \mathbf{b_2^*}\mathbf{b_1} & \mathbf{b_2^*}\mathbf{b_2} & \mathbf{b_2^*}\mathbf{b_3} \\ \mathbf{b_3^*}\mathbf{b_1} & \mathbf{b_3^*}\mathbf{b_2} & \mathbf{b_3^*}\mathbf{b_3} \end{pmatrix} = \begin{pmatrix} 1 & 0 & 0 \\ 0 & 1 & 0 \\ 0 & 0 & 1 \end{pmatrix}. \tag{3.15}$$

Therefore $\mathbf{Q} \cdot \mathbf{r_j} = hx_j + ky_j + lz_j$.

Interference appears when we calculate the bulk structure factor, F_{bulk}, by summing over all the unit cells in the crystal. To do so we have to multiply the sums with a phase factor that propagates the structure factor of one unit cell to all the unit cells in the crystal.

$$F_{bulk}(\mathbf{Q}) = F_{uc}(\mathbf{Q}) \sum_{u=1}^{U} \exp(2\pi i u \mathbf{Q} \cdot \mathbf{b_1}) \sum_{v=1}^{V} \exp(2\pi i v \mathbf{Q} \cdot \mathbf{b_2}) \sum_{w=1}^{W} \exp(2\pi i w \mathbf{Q} \cdot \mathbf{b_3}) \tag{3.16}$$

3. Experimental details

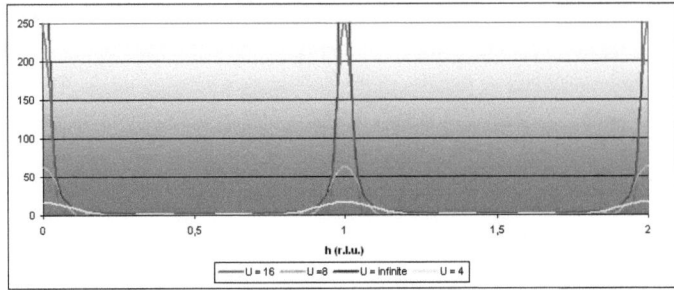

Fig. 3.5: $|S_U|^2$ as a function of h. Peaks appear at integer values of h. Peak height is U^2 and peak width at half maximum is $2\pi/U$.

The projection of $\mathbf{Q}$ on $\mathbf{b_1}$ is equal to h (according to equations 3.14 and 3.15) and the sums in equation 3.16 represent geometric series and can therefore be rewritten as shown here exemplarily for the first sum, S_U:

$$S_U = \sum_{u=1}^{U} \exp(2\pi i u \mathbf{Q} \cdot \mathbf{b_1}) = \frac{1 - \exp(2\pi i U h)}{1 - \exp(2\pi i h)} \; ; \lim_{U \to \infty} S_U = \frac{1}{1 - \exp(2\pi i h)} \; . \quad (3.17)$$

S_U represents a simplified version of the scattering amplitude for a one dimensional crystal with U unit cells. The corresponding intensity would be given by the square modulus of S_U.

$$|S_U|^2 = \frac{\sin^2(\pi U h)}{\sin^2(\pi h)} \; ; \lim_{U \to \infty} |S_U|^2 = \frac{1}{\sin^2(\pi h)} \quad (3.18)$$

The appearance of $|S_U|^2$ as a function of h for some values of U is shown in Figure 3.5. Peaks appear when h takes integer values. In 3D crystallography the sums are taken from $-\infty$ to $+\infty$ and can therefore be replaced by delta–functions of the form: $\delta(h-u); u \in \mathbf{Z}$. Scattering intensity is only observed at "Bragg–peaks" when h, k, and l take integer values and the peak intensities are proportional to $|F_{uc}|^2$. This leads to the definition of the reciprocal lattice and the "Laue"–conditions. In 2D crystallography, and the surface of a crystal can be regarded as a 2D lattice, leading to scattering intensity at integer h and k values for all l's (taking l as the momentum transfer perpendicular to the surface plane). To calculate structure factors for non–integer l values delta–functions can be used in the $\mathbf{b_1}$ and the $\mathbf{b_2}$ directions, while a semi-infinite sum is used in the $\mathbf{b_3}$ direction. It is this semi–infinite sum that results in scattering intensity at non–integer l values. A simple equation

3. Experimental details

can now be formulated to calculate bulk structure factors, F_{bulk}, that describe scattering from a semi–infinite crystal (quasi infinite in the $\mathbf{b_1}$ and the $\mathbf{b_2}$ directions and semi-infinite in the $\mathbf{b_3}$ direction). Robinson (1986) [90] invented the name Crystal Truncation Rod (CTR) for these rod like features in reciprocal space perpendicular to the crystal surface.

$$F_{bulk}(\mathbf{Q}) = F_{uc}(\mathbf{Q})\delta(h-u)\delta(k-v) \sum_{w=-\infty}^{0} \exp(2\pi i w l)\exp(w\alpha)\,; u,v \in \mathbf{Z} \qquad (3.19)$$

The attenuation factor, α, is included because with increasing depth the contribution of unit cells to scattering decreases and in reality only a finite number of unit cells contribute to the CTR.

However, equation 3.19 does not yet include any expression to specifically address effects of surface atoms relaxed from their bulk positions or structured water molecules above the surface on the scattering intensity. To do so surface structure factors, F_{surf}, are calculated analogously to the unit cell structure factors.

$$F_{surf}(\mathbf{Q}) = \sum_j f_j(\mathbf{Q})\theta_j \exp(2\pi i \mathbf{Q}\cdot \mathbf{r_j})\exp\left(-\frac{B_j Q^2}{4}\right). \qquad (3.20)$$

The surface structure factor is calculated for a surface unit cell. This includes atoms from the top molecular layers of the crystal that might be affected by structural relaxation and in case adjacent water layers or anything else that is ordered and sits on top of the crystal. As surface sites may not all be fully occupied, the additional site occupancy factor, θ_j, is included. The coherent sum of F_{surf} and F_{bulk}, F_{sum}, is directly related to the scattered intensity. A scaling factor, S, and a roughness factor, R, are included into the calculation of F_{sum}.

$$F_{sum}(\mathbf{Q}) = SR(F_{surf}(\mathbf{Q}) + F_{bulk}(\mathbf{Q})) \qquad (3.21)$$

Comparison of measured and calculated structure factors allows to refine positions, Debye–Waller factors, and occupancies of atoms in the surface unit cell by a least square fitting procedure.

3.6.2. Surface unit cell

To facilitate surface diffraction measurements and data analysis at the calcite (104)–face a new pseudo–orthogonal surface unit cell has been defined based on the hexagonal unit cell described in section 2.1.2. An approach published by Trainor et al. (2002) [91] is

3. Experimental details

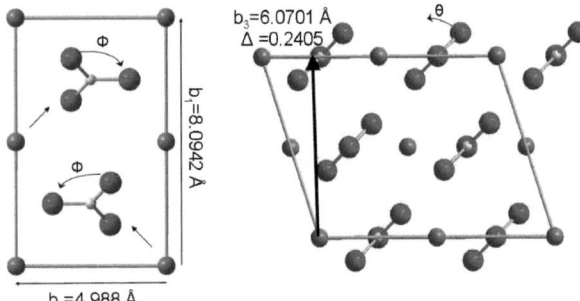

Fig. 3.6: The rectangular unit mesh at the calcite (104)–face shown on the left side is used to define basis vectors $\mathbf{b_1}$ and $\mathbf{b_2}$ of the pseudo–orthogonal calcite surface unit cell. The right side shows the projection of the calcite structure along the $\mathbf{b_2}$ direction. The orange frame depicts a monoclinic unit cell. The thick black arrow depicts the basis vector $\mathbf{b_3}$ which is perpendicular to the surface. The length of $\mathbf{b_3}$ is twice the layer thickness, d_{104}, of the calcite (104)–plane. Δ is the offset between the corner of the monoclinic cell and the corner of the pseudo–orthogonal cell along $\mathbf{b_1}$ in fractional length units. To maintain glideplane symmetry along the $\mathbf{b_1}$ direction during the structure refinement translations in $\mathbf{b_2}$ direction and carbonate rotation ϕ are defined as shown on the left side of the Figure. Definition of the carbonate rotation θ is shown on the right hand side.

used to define the surface unit cell and to calculate structure factors according to that unit cell. The rectangular unit mesh at the calcite (104)–face that is used to define basis vectors $\mathbf{b_1}$ and $\mathbf{b_2}$ of the pseudo–orthogonal calcite surface unit cell is shown on the left hand side in Figure 3.6. The right side shows the projection along the $\mathbf{b_2}$ direction. The orange frame depicts a monoclinic unit cell. The thick black arrow depicts the new basis vector $\mathbf{b_3}$. The length of $\mathbf{b_3}$ is 6.0701 Å, twice the layer thickness, d_{104}, of the calcite (104)–plane. $\Delta = 0.2405$ is the offset between the corner of the monoclinic cell and the corner of the pseudo–orthogonal cell along $\mathbf{b_1}$ in fractional length units. According to the definition of the new pseudo–orthogonal surface unit cell, the calcite (104)–face becomes the (001)–face. The consequence of using this unit cell that has no translational symmetry along the $\mathbf{b_3}$ direction is that the $l = 0$ level of all CTRs is at the crystal surface. However, the Bragg–peaks on CTRs with $h \neq 0$ do no longer appear at integer values of l, but at values of $l = n - \Delta \cdot h$ with $n \in \mathbf{Z}$. The fractional coordinates of the atoms in the surface

3. Experimental details

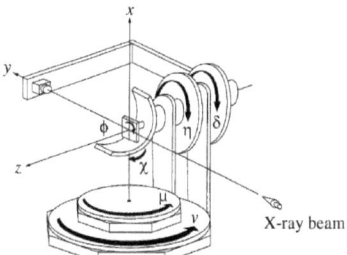

Fig. 3.7: Definitions of angles on a six (4+2) circle diffractometer and the diffractometer coordinate system [92].

unit cell can be found in the bulk–file and fit–file attached in appendix A. Having the b_3 direction perpendicular to the surface facilitates the data analysis. The displacement of atoms along this direction now corresponds to the surface normal direction, it is less correlated to lateral displacement and is easy to imagine.

3.6.3. Surface diffraction measurements

Surface diffraction measurements at the GeoSoilEnviroCARS undulator beamline, 13IDC, at the Advanced Photon Source (APS) in Argonne, have been performed on a Newport Kappa six (4+2) circle diffractometer. Definitions of angles used in this description are shown in Figure 3.7. For measurements of the specular (00L) CTRs the angle between the sample surface normal and the horizontal (yz)–plane, NAZ, is fixed at 12°, μ is fixed at 0°, and η is defined as $\eta = \delta/2$. Off–specular CTRs are measured with the angle between the sample surface and the x–ray beam, α, fixed at 2°, NAZ fixed at 12°, and η defined as $\eta = \delta/2$.

The beamline optics consist of a liquid nitrogen cooled Si(111) double crystal monochromator and rhodium coated horizontal and vertical focusing mirrors. The incident x–ray energy is tuned to 16 keV on the third harmonic of the undulator. This corresponds to an x–ray wavelength of 0.7749 Å. The beam is focused to a size of 100 × 1000 μm^2 (horizontal × vertical). The diffracted beam is detected on a PILATUS 2D pixel array detector with 195 × 487 pixels (vertical × horizontal). Using the array detector instead of a traditional

63

3. Experimental details

x–ray scintillator to scan rocking curves reduces the data acquisition time remarkably. Extraction of structure factors from the PILATUS images is performed using the PDS software package. PDS (Python Data Shell) is an open source software, part of the tdl project [93]. Integrated intensities are corrected for polarization by the polarization factor C_P and for the variation in the intercept between Ewald–sphere and CTR as a function of $\mathbf{Q}$ by the factor C_I, according to the method published by Schlepütz et al. (2005) [94] assuming 100% polarization.

$$C_P = 1 - \cos^2 \delta \sin^2 \nu \tag{3.22}$$

$$C_I = \cos \delta \sin \beta \tag{3.23}$$

where β is the angle between the sample surface and the diffracted beam. Further corrections concern the area of the beam footprint on the sample, A, and the part of the beam that hits the sample surface, S. The integrated intensities, I, need to be corrected for the incident intensity, I_0. Structure factor magnitudes, $|F|$, can be calculated as:

$$|F| = \sqrt{\frac{I}{I_0} \cdot \frac{C_I}{C_P A S}} \tag{3.24}$$

Measurements have been conducted on freshly cleaved calcite single crystals var. Iceland spar from Mexico Chihuahua purchased form Ward's natural science. Directly after cleavage an optically clear crystal platelet is mounted on the sample holder which is then mounted onto the diffractometer and covered with a mylar semi–spherical hood that can be flushed with helium to avoid contamination with adventitious carbon. An initial alignment and scanning of the crystal is performed to assure perfect crystallinity. After that the crystal is covered with an adjustable sleeve with an upper 8 μm thick Kapton foil, which is sealed with an O–ring and constitutes the top cover of the liquid sample cell used to measure the 3D structure of the mineral water interface in situ through a thin film of solution ($\sim 1\,\mu$m). A schematic drawing of the sample cell used for in situ surface diffraction measurements on mineral solution interfaces is shown in Figure 3.8. While the sample holder with the crystal on top is fixed on the diffractometer, the outer sleeve carrying the Kapton foil can be moved up and down with a motor to tighten the Kapton foil over the sample. The space between sample holder and Kapton can be flushed with solution. Before each measurement the sleeve is raised thereby lifting the Kapton foil off the sample and the sample is flushed with solution several times and allowed to equilibrate for about 15

3. Experimental details

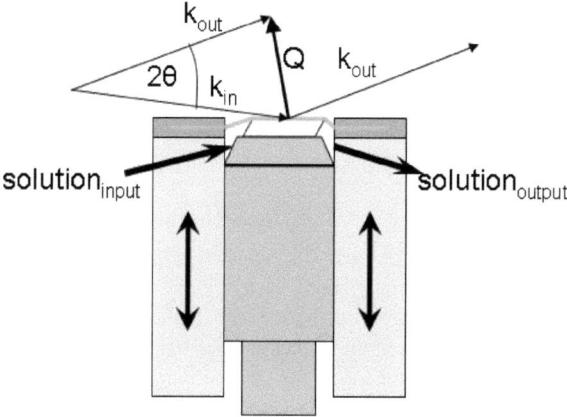

Fig. 3.8: Schematic drawing of the sample cell used for in situ surface diffraction measurements on mineral solution interfaces. The sample holder with the crystal on top (cyan and white) is fixed on the diffractometer. The outer sleeve covered with Kapton foil (gray and yellow) can be moved up and down by a motor to tighten the Kapton foil over the sample (indicated by arrows). The space between sample holder and Kapton can be flushed with solution. The geometry relation between incoming and outgoing wave vectors and the scattering vector **Q** is indicated.

min. Then the Kapton is lowered while pumping off solution from the sample cell using a peristaltic pump. When the Kapton is close to the surface and the rim of the O–ring sealing the Kapton foil (gray part in Figure 3.8) is below the crystal surface, the rest of the solution is sucked out of the cell. The ideal solution film thickness is reached when Newton rings (visible interference phenomena between visible light reflected from the sample surface and the Kapton foil) appear on the sample surface. The film thickness, d, should be of about the same order of magnitude as the wavelength of visible light, λ ($d = n/2\,\lambda$).

Solutions used for the measurements have been pre–equilibrated with the method described above (cp. section 3.4). Seven datasets (A–G) consisting of nine crystal truncation rods (only seven in dataset G) have been collected under different conditions:

A : Preliminary measurements under helium atmosphere mainly used to check the quality

3. Experimental details

of the crystal. The Data has not yet been analyzed in detail.
Measured CTRs: (00L), (01L), (02L), (20L), ($\bar{2}$0L), (21L), ($\bar{2}$1L), (13L), ($\bar{1}$3L).

B : Solution containing 0.097 M NaCl and 0.003 M NaOH in equilibrium with calcite and air (atmospheric CO_2), expected pH: 8.6, expected zetapotential: $\sim$ 0 mV.
Measured CTRs: (00L), (01L), (02L), (20L), ($\bar{2}$0L), (21L), ($\bar{2}$1L), (13L), ($\bar{1}$3L).

C : Solution containing 0.03 M NaCl and 0.07 M HCl in equilibrium with calcite and air (atmospheric CO_2), expected pH: 7.5, expected zetapotential: $\sim$ 9 mV.
Measured CTRs: (00L), (01L), (02L), (20L), ($\bar{2}$0L), (21L), ($\bar{2}$1L), (13L), ($\bar{1}$3L).

D : Non–equilibrium solution containing 0.001 M NaOH, 0.01 M $CaCl_2$, and 0.01 M NaCl, expected pH: 10.9, expected zetapotential: $\sim$ 8 mV.
Measured CTRs: (00L), (01L), (02L), (20L), ($\bar{2}$0L), (21L), ($\bar{2}$1L), (13L), ($\bar{1}$3L).

E : Non–equilibrium solution containing 0.0005 M NaOH, 0.01 M Na_2CO_3, and 0.01 M NaCl, expected pH: 11.1, expected zetapotential: $\sim$ -25 mV.
Measured CTRs: (00L), (01L), (02L), (20L), ($\bar{2}$0L), (21L), ($\bar{2}$1L), (13L), ($\bar{1}$3L).

F : Solution containing 0.002 M NaOH in equilibrium with calcite and air (atmospheric CO_2), expected pH: 8.6, expected zetapotential: $\sim$ 0 mV.
Measured CTRs: (00L), (01L), (02L), (20L), ($\bar{2}$0L), (21L), ($\bar{2}$1L), (13L), ($\bar{1}$3L).

G : Solution containing 0.03 M NaCl and 0.07 M HCl in equilibrium with calcite and CO_2, expected pH: 5.8, expected zetapotential: $\sim$ 12 mV.
Measured CTRs: (00L), (01L), (02L), (20L), ($\bar{2}$0L), (21L), ($\bar{2}$1L).

For datasets A – D the (00L) CTR has been measured twice to check for beam damage on the sample and to estimate systematic errors of the measurements. No indication for beam damage could be observed. The structure factors obtained in consecutive measurements are equal within measurement uncertainty.

A modified version of the ROD program [95] is used for the structure refinement. The definition of the initial structure and the fitting parameters used for the refinement can be found in the bulk–file and fit–file attached in appendix A. The initial structure resembles bulk terminated calcite with four layers of water on top. As water hydrogens have hardly any electron density around them, they are invisible in x–ray scattering. Therefore, in the

3. Experimental details

fit file water is substituted by oxygen atoms. Three translational parameters along $\mathbf{b_1}$, $\mathbf{b_2}$, and $\mathbf{b_3}$, and an isotropic Debye–Waller factor are fitted for all atoms in the top two calcite monolayers and the water layers. Carbonate ions in the top two calcite monolayers are defined as groups. Atoms of the planar carbonate ions have identical translational parameters. They are translated parallel to each other. They can be rotated around the $\mathbf{b_2}$ vector by the angle θ (see Figure 3.6) and around the molecule normal by the angle ϕ. Center of rotation is the central carbon atom.

Breaking of the glideplane symmetry along the $\mathbf{b_1}$ direction would cause scattering intensity on the $(\bar{1}0L)$ rod, the $(10L)$ rod, the $(\bar{3}0L)$ the $(30L)$ rod and so on. It is known that at the calcite (104)–surface glideplane symmetry is maintained even upon relaxation in contact to solution [49]. How rotational angles and translation in the $\mathbf{b_2}$ direction are defined, in order to maintain the glideplane symmetry along the $\mathbf{b_1}$ direction during structure refinement, is indicated by arrows in Figure 3.6.

Bond valence calculations [96] are used to check the plausibility of the refined surface structures. In a reasonable structure the bond valence sum of the surface calcium atoms should not deviate too much from two. Bond valence parameters for Ca – O bonds are: $R_0 = 1.967$ and b = 0.37 [96].

3.7. Surface complexation modeling

A Basic Stern model has been implemented in order to calculate the surface speciation based on observed zeta potentials and consistent with the results of surface diffraction measurements. The principle of a Basic Stern model is illustrated in Figure 3.9. The potential, ψ (V), is plotted as a function of the distance, x (m). Charge, σ (C/m^2), can be placed in the 0– and the b–plane (σ_0 and σ_b, respectively) the two planes that confine the Stern layer. Beyond the b–plane the diffuse layer begins characterized by an exponential decrease in potential according to the Gouy–Chapman equation. As all experiments have been performed in NaCl solutions the equations for a monovalent electrolyte are used. A slip plane is included in the diffuse layer in order to fit zetapotentials. The slip plane distance, d_S, (distance between b–plane and slip plane) is fit independently for datasets at different ionic strengths.

Two principle surface sites are considered: >CaOH and >CO$_3$H. Site density is 4.95 nm^{-2} each. In the 0–plane only protonation / deprotonation reactions are considered. All

3. Experimental details

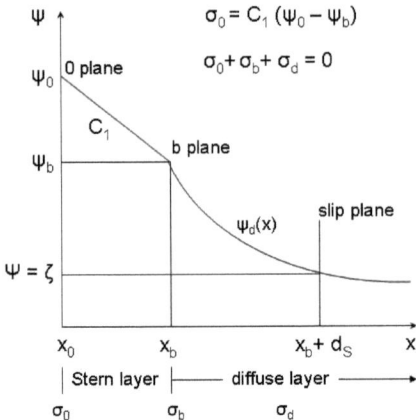

Fig. 3.9: The Basic Stern model. Shown is potential, ψ (V), against distance, x (m). Charge, σ (C/m^2), can be placed in the 0– and the b–plane.(after [97])

other surface–ion interactions are assumed to take place in the b–plane. Type I components (known concentration) in the model are: Ca^{2+}, CO_2, Na^+, Cl^-, $>CaOH^{(x-1)}$, and $>CO_3H^{(1-x)}$. H^+ is defined as type III component (known activity). The A–matrix that defines how the speciation is calculated is listed in Table 3.1. The inner Helmholtz capacitance, C_1, and four slip plane distances ($x_1 - x_4$), the two reaction constants of the protonation / deprotonation reactions (K_1 and K_2) and the five ion binding constants (IB) that are defined in the A–matrix are the only adjustable parameters.

The UCODE inverse modeling code [98] is used for the adjustment of surface complexation model parameters. A modified version of the Fiteql 2.0 [99] software is used with UCODE for equilibrium calculations. The activity coefficient, γ, is calculated according to the Debye–Hückel equation assuming constant ionic strength for samples of one dataset. It is used as indicated in the last column of the A–matrix. Fourteen datasets are included in the fitting procedure. How the data has been obtained is described in the section on zetapotential measurements (3.5). Six pH series from electrophoretic zetapotential measurements at equilibrium with N_2, air, and CO_2 two series at each CO_2 partial pressure are included. Five streaming potential titrations with 0.01 M NaCl as background electrolyte at various levels of calcium concentration (0 mM, 0.1 mM, 0.5 mM, 1 mM, and 5 mM) and

3. Experimental details

one streaming potential titration with 5 mM carbonate in solution are included. The last two datasets explore the influence of ionic strength; NaCl concentration: 0.001 M and 0.05 M.

3. Experimental details

Tab. 3.1: The A-matrix of the Basic Stern model for calcite.

A–Matrix	Ca^{2+}	CO_2	Na^+	Cl^-	H^+	$>CaOH$	$>CO_3H$	ψ_0	ψ_b	$\log_{10} K$	γ
CO_3^{2-}	0	1	0	0	-2	0	0	0	0	-18.15 $(= K_{CO_3})$	-4
HCO_3^-	0	1	0	0	-1	0	0	0	0	-7.82 $(= K_{HCO_3})$	-1
$CaHCO_3^+$	1	1	0	0	-1	0	0	0	0	-17.04	+3
$CaCO_{3(aq)}^0$	1	1	0	0	-2	0	0	0	0	-14.93	+4
$CaOH^+$	1	0	0	0	-1	0	0	0	0	-12.70	+3
$>CaOH^{(x-1)}$	0	0	0	0	0	1	0	0	0	0.00	0
$>CaOH^{(x-1)} \cdots Na^+$	0	0	1	0	0	1	0	0	1	$IB(Na^+)$	+1
$>CaOH^{(x-1)} \cdots Ca^{2+}$	1	0	0	0	0	1	0	0	2	$IB(Ca^{2+})$	+4
$>CaOH_2^x$	0	0	0	0	1	1	0	1	0	K_1	0
$>CaOH_2^x \cdots Cl^-$	0	0	0	1	1	1	0	1	-1	$K_1 + IB(Cl^-)$	+1
$>CaOH_2^x \cdots HCO_3^-$	0	1	0	0	0	1	0	1	-1	$K_1 + K_{HCO_3} +$ $IB(HCO_3^-)$	0
$>CaOH_2^x \cdots CO_3^{2-}$	0	1	0	0	-1	1	0	1	-2	$K_1 + K_{CO_3} +$ $IB(CO_3^{2-})$	0
$>CO_3H^{(1-x)}$	0	0	0	0	0	0	1	0	0	0.00	0
$>CO_3H^{(1-x)} \cdots Cl^-$	0	0	0	1	0	0	1	0	-1	$IB(Cl^-)$	+1
$>CO_3H^{(1-x)} \cdots HCO_3^-$	0	1	0	0	-1	0	1	0	-1	$K_{HCO_3} + IB(HCO_3^-)$	0
$>CO_3H^{(1-x)} \cdots CO_3^{2-}$	0	1	0	0	-2	0	1	0	-2	$K_{CO_3} + IB(CO_3^{2-})$	0
$>CO_3^{-x}$	0	0	0	0	-1	0	1	-1	0	K_2	0
$>CO_3^{-x} \cdots Na^+$	0	0	1	0	-1	0	1	-1	1	$K_2 + IB(Na^+)$	+1
$>CO_3^{-x} \cdots Ca^{2+}$	1	0	0	0	-1	0	1	-1	2	$K_2 + IB(Ca^{2+})$	+4

3.8. Adsorption experiments

3.8.1. The experimental setup and procedure

Batch type adsorption experiments have been carried out using an experimental setup very similar to the one used for calcite equilibration before zetapotential or surface diffraction measurements. pH is adjusted by variation of CO_2 partial pressure (at atmospheric pressure) no acid or base is added. A scheme of the experimental setup is shown in 3.3. Adsorption has been investigated at four pH values: at pH 6 equilibrating the calcite suspension with a pure CO_2 gas phase, pH 7.2 using 2 % CO_2 in N_2, pH 8.3 using air (360 ppm CO_2), and pH 9.4 using N_2 with a CO_2 content of about 6 ppm. For the experiments with air the gas bottle in Figure 3.3, is replaced by an electric air pump. The advantage of this approach of pH adjustment is that within the whole pH range studied the ionic strength of the solutions varies by 30 % only (0.1—0.13 M). Compared to pH adjustment by strong acids or bases at constant CO_2 partial pressure, model calculations show that ionic strength would vary by 300 % in the same pH region (0.1—0.3 M). It is important that the ionic strength is constant to ensure that the adsorption data at the different pH values are directly comparable, because ionic strength has a major influence on adsorption, especially outer–sphere adsorption. Calcite powder Merck calcium carbonate suprapur is used, which was shown by XRD measurements to consist only of calcite ($\pm$ 1%). XPS analyses revealed possible small amounts of organic carbon at the crystal surfaces; about 20 % of the C 1s spectrum can be allocated to organic carbon. This is small compared to amounts of adventitious carbon at calcite surfaces in contact with the atmosphere reported in the literature [100]. The specific surface of the calcite powder has been measured by N_2–BET to be 1.28 $\pm$0.11 m^2/g. SEM images of the calcite powder show that calcite crystals have a size distribution between about 5 and 30 μm in diameter. From these images (an example is shown on the left side in Figure 2.3 and on the left side in Figure 3.13), it seems valid to assume that the calcite crystals consist mainly of (104)–rhombohedra. This implies that it is reasonable to assume a 1:1 ratio of >Ca and >CO3 surface sites with a site density of 4.95 nm^{-2}, as usually done in calcite surface complexation models [60, 62] (cp. section 2.2.1). Before every experiment, 0.1 M NaCl solution is allowed to equilibrate with calcite and the gas phase for about 24 h. pH measurements are performed to verify that the calculated equilibrium pH is reached during the pre-equilibration period. Then 25 mL of the equilibrated solution is added to 500.5 $\pm$ 0.5 mg calcite. An aliquot of 5 mM

3. Experimental details

pH 2.5 NpO_2^+ stock solution is added to reach the desired initial $^{237}NpO_2^+$ concentration between 0.4 μM and 30 μM. In the experiment at pH 8.3, the concentration range has been extended to 40 μM NpO_2^+. The suspensions are continuously percolated with gas bubbles during the experiment. Experiments on adsorption kinetics, performed with an initial NpO_2^+ concentration of 2 μM at pH 8.3, show that after 2 days the adsorption slows down remarkably. Therefore, for the determination of adsorption isotherms 72 h of reaction time are considered. At the end of an experiment the sample tube is opened, the pH is measured, and, after sedimentation of the calcite powder a sample of the supernatant is analyzed by LSCto quantify Np-237. Separation by sedimentation works well and fast for well crystalline material like calcite with a particle size >5 μm. It has therefore been preferred over centrifugation or filtration, which promote pH changes due to changes in CO_2 partial pressure. At each pH value, 6 to 12 experiments have been carried out. Adsorption to container walls has been investigated in blank experiments and has been found to be negligible. Model calculations show that all the experimental solutions are undersaturated with respect to the relevant solid neptunyl carbonate phase, $NaNpO_2CO_3 \cdot 3.5\ H_2O$ (s, fr). For the high concentration at the higher pH values, the solubility limit is almost reached. Adsorption isotherms show no indication for precipitation of a neptunyl carbonate phase.

3.8.2. Low temperature (15 K) EXAFS measurements

The samples prepared for EXAFS measurements at the European Synchrotron Radiation Facility (ESRF) in Grenoble have an initial neptunyl concentration of 40 μM, solution pH is 8.3. Two samples (Np-O1 and Np-O2) have been allowed to react for three months, and two samples (Np-Y1 and Np-Y2) for 48 hours. After the reaction time suspensions are allowed to sediment and the supernatant is removed. The reacted calcite powder is filled as a wet paste into the SH01B-cryoholder, designed by the Rossendorf beamline (ROBL) staff for low temperature EXAFS measurements of low activity radioactive samples. After sealing the sample holders the samples are frozen in liquid nitrogen, in order to stop the reaction. During transport to the ESRF in Grenoble, samples are cooled in dry ice. EXAFS measurements at the Np L_3-edge (17610 eV) are performed at the Rossendorf beamline (ROBL) [101] at the ESRF in fluorescence mode at low temperature (15 K) in a Helium cryostat inside a glovebox. Parallel measurement of a zirconium foil (Zr K-edge: 17998 eV) in transmission mode enabled energy calibration of the spectra. Fifteen spectra of

3. Experimental details

sample Np-O2 have been measured and 16 spectra of sample Np-Y1. Dead time correction is performed using the SIXpack [102] software. Spectra are merged using Athena [103]. Background subtraction, μ_0-fitting, and conversion to k-space are performed with WinXAS 3.1 [104]. As no significant difference between the young (Np-Y1) and the old sample (Np-O2) is observed the structure has been analyzed from the spectrum with the better signal to noise ratio, which is Np-O2. For calculation of theoretical backscattering and phase shift functions and for the data fitting, the Artemis [103] software is used. The extracted EXAFS signal is used in a k-range from 2.0 Å^{-1} to 10.2 Å^{-1}. Using a larger k–range caused significant noise contribution to the R-space spectrum. Hanning windows are used in the Fourier transformations. The axial oxygen scattering path is used to correct for phase shift. Fits are performed in R-space in the 1 to 5 Å R–range, with k-weights 1, 2, and 3. Bond-valence calculations [96] are used to check the plausibility of the of the obtained coordination environment around the central neptunium atom. Bond valence parameters for Np – O bonds are: $R_0 = 2.035$ and $b = 0.422$ [105].

3.9. Coprecipitation experiments

3.9.1. Mixed flow reactor (MFR) experiments

Coprecipitation and calcite growth experiments are conducted in a mixed flow reactor (MFR) at room temperature. This way of synthesis has the advantage that in the reactor steady state conditions are maintained, the solution composition remains constant, and the doped calcite can grow homogeneously onto seed crystals. Homogeneous partition coefficients and growth rates can be determined from the experiment.

Figure 3.10 illustrates the experimental setup of MFR experiments. Three solutions (A 1–3), the first containing carbonate, the second neptunium (or uranium), and the third calcium, are pumped by a peristaltic pump (B) into the MFR (C). All solutions contain the background electrolyte, NaCl. A magnetic stir bar is inside the MFR. The suspension of seed crystals inside the reactor is agitated by the magnetic stirrer (D). The whole reactor is located in a water bath (E) which is temperature controlled by a heating coil connected to a thermostat. In this way the temperature can be kept constant ($\pm$ 0.1°C) over several weeks. A discharge valve (F) is placed between the output of the reactor and the waste container (G) for taking samples.

3. Experimental details

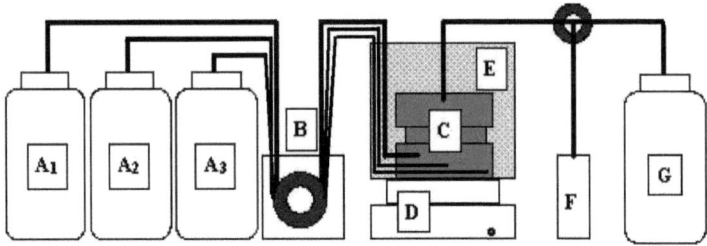

Fig. 3.10: Setup for mixed flow reactor experiments

Figure 3.11 shows a detailed sketch of the MFR. The three connections for the different solution input tubes are shown. (1) is the magnetic stirring bar. It is suspended by a Teflon construction that is fixed at the reactor walls to avoid grinding effects. (2) is the filter. The outgoing solution is filtered by a 0.45 μm pore size filter paper, to keep the seed crystals from leaving the reactor. The output tube is connected at the top of the reactor. The reactor has a reaction volume of approximately 45 mL. Pictures of the experimental setup and the sampling procedure are shown in Figure 3.12.

The Stock solutions are prepared from 18.2 $M\Omega$ MilliQ water and Merck p.a. chemicals. Neptunium(V) has been taken from stock solutions of NpO_2^+ dissolved in perchloric acid (pH 2–3). Merck calcium carbonate suprapur is used as seed crystals. Before use the composition of the seed crystals has been analyzed for dissolved organic carbon and the surface has been investigated by XPS measurements to preclude organic contaminations. Comparison of the measured XPS spectra to results reported by Stipp and Hochella (1991) [100] show that there is no significant contamination with organic matter. The specific surface area of the seed crystals has been measured by BET; it is $1.28 \pm 0.11\,\mathrm{m^2/g}$.

The diameter of seed crystals should be larger than 5–10 μm to minimize the boundary layer effect and to ensure surface controlled precipitation kinetics [36, 106] (cp. section 2.3). SEM images of crystal seeds show that the particle diameters generally range from 5 μm to 30 μm. Attempts to sieve out a 11–15 μm fraction have not shown any significant effect. SEM images of crystal seeds before and after sieving are shown in Figure 3.13. The BET surface before and after sieving is within error constant at 1.28 $\mathrm{m^2/g}$. Therefore, after some initial experiments with the sieved calcite untreated Merck calcium carbonate suprapur has been used as crystal seeds.

3. Experimental details

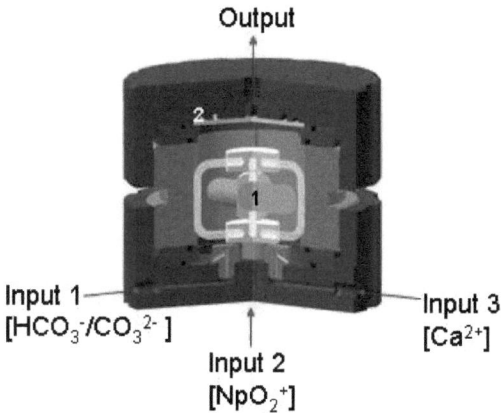

Fig. 3.11: Detailed sketch of the MFR

Stock solutions are sampled before the experiment. During the experiment the reactor output is sampled daily. Calcium and neptunium concentrations in the samples are measured by ICP-MS. Calcium concentrations in the inactive experiments are measured by ICP-OES. The flow rate is determined with each sampling by clocking the time of sampling and measuring the sample volume. A slight decrease ($< 5\ \%$) in flow rate during the run of an experiment is typical for peristaltic pumps due to fatigue of the pump tubes. Thirteen mL sample are typically collected. Ten mL are acidified with 10 μL concentrated hydrochloric acid and stored for later analysis and the remaining 3 mL are used to measure pH. Assuming perfect mixture in the reactor, the solution composition of the outflowing solution is equal to the homogeneous solution composition in the reactor.

From the concentration entering the reactor (c_{in}), the concentration leaving the reactor (c_{out}), and the flow rate (F) the crystal growth rate (R) can be calculated. Usually the crystal growth rate is given normalized by the crystal surface area present in the reactor (A). The surface area is assumed to remain constant during the whole experiment.

$$R = \frac{F \cdot (c_{in} - c_{out})}{A} \qquad (3.25)$$

3. Experimental details

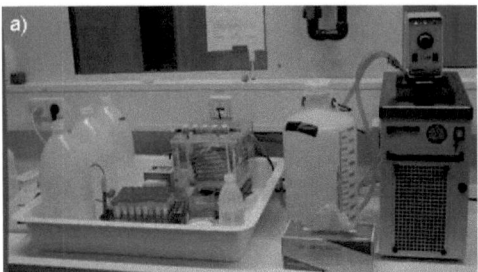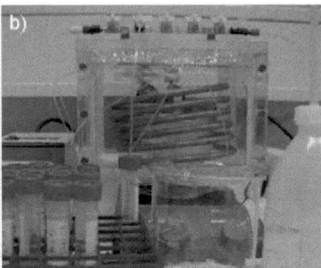

Fig. 3.12: Photographs of the whole experimental setup (a), and of the sampling procedure, with the MFR in the water quench in the background (b).

By inserting calcium concentrations into this equation the calcite growth rate (R_{Ca}) can be obtained, inserting neptunium concentrations yields the rate of neptunium incorporation (R_{Np}). Typically during the course of the experiment both rates decrease and attain a more or less constant value, the steady state value, after a few days. The faster rates at the beginning of the experiment can be explained by saturation of high active sites at the seed crystal surfaces or by adsorption phenomena at reactor- or tube walls. The mean reaction rate during steady state conditions ($\overline{R}$) can be calculated either by taking the average of the reaction rates or by calculating it from the mean flow rate during steady state conditions ($\overline{F}$) and the mean output concentration during steady state conditions ($\overline{c_{out}}$). Multiplying the mean reaction rate at steady state conditions ($\overline{R}$) by the time during which the system was kept in steady state and the crystal surface area present in the reactor (which is assumed to remain constant during the whole experiment), we obtain the total amount of calcium or neptunium cations precipitated during steady state. As the steady state duration time and the surface area are the same for neptunium and calcium, molar fractions in the grown solid solution can be calculated directly from the mean steady state reaction rates.

$$X_{Np} = \frac{\overline{R_{Np}}}{\overline{R_{Ca}} + \overline{R_{Np}}} \qquad (3.26)$$

$$X_{Ca} = \frac{\overline{R_{Ca}}}{\overline{R_{Ca}} + \overline{R_{Np}}} \qquad (3.27)$$

3. Experimental details

Fig. 3.13: SEM images of calcite crystals used as crystal seeds in MFR experiments. Left image: untreated Merck calcium carbonate suprapur. Right image: sieved fraction (11-15 μm) of Merck calcium carbonate suprapur.

With these molar fractions and the mean steady state output concentrations $\overline{[Ca]}$ and $\overline{[Np]}$, the empirical partition coefficient D can be calculated.

$$D = \frac{X_{Np}/X_{Ca}}{\overline{[Np]}/\overline{[Ca]}} \quad (3.28)$$

This partition coefficient sets the ratio of the molar fractions in the solid solution in proportion to the ratio of the concentrations in the aqueous solution. It is meant to describe the affinity of a trace element for incorporation into a host mineral. Values smaller than one indicate that the trace element is relatively enriched in the aqueous solution, for values larger than one the trace element is enriched in the solid solution.

Molar fractions in equation 3.28 can be replaced by the mean steady state reaction rates.

$$D = \frac{\overline{R_{Np}/R_{Ca}}}{\overline{[Np]}/\overline{[Ca]}} \quad (3.29)$$

After inserting the equation for the reaction rate, D becomes:

$$D = \frac{\frac{[Np]_{in}}{[Np]} - 1}{\frac{[Ca]_{in}}{[Ca]} - 1} \quad (3.30)$$

These calculation steps are important because in this way the partition coefficients becomes independent of the flow rate and the surface area and thereby the calculated error becomes much smaller.

3. Experimental details

The pH values of the stock solutions have been measured regularly. As the solutions in the reservoirs are not in equilibrium with atmospheric CO_2, a change in pH, especially of the carbonate solution, can be observed. When the pH of the input solution is observed to change too much during the experiment (> 0.2 pH units), the carbonate solution is replaced. The pH and SI in the reactor (the output solution) are monitored, in order to be able to characterize the steady state conditions. The total inorganic carbon concentration, TIC, in the reactor is calculated from the input concentration and the change in calcium concentration, assuming that the total carbon concentration decreases by the same amount as the calcium concentration during precipitation. Typically the duration of one experiment is 2–3 weeks. Doped calcite product has been taken out of the reactor after some of the experiments and prepared for EXAFS measurements.

The solution compositions needed to obtain the desired supersaturation and pH are calculated with PhreeqC and the Nagra/PSI thermodynamic database [32]. The supersaturation of the input solutions has been chosen not to exceed the supersaturation found by Tai et al. (1999) [68] to be metastable towards homogeneous nucleation, which corresponds to a saturation index of about one. For this supersaturation, formation of the precursor phase vaterite can be excluded and the influence of diffusion on growth rate is negligible [36].

Error propagation is used to determine the errors for all the values. The initial errors of the concentrations and the specific surface are calculated from analysis of standard samples. Initial errors for calculating the flow rate error are estimated to be: ± 0.1 mL for the volume measurement and ± 5 s for clocking the sampling time.

Sixteen MFR experiments have been carried out: four to study precipitation kinetics of pure calcite in the MFR, Cc1 - Cc4, eleven to study neptunyl(V) coprecipitation with calcite, Np1 - Np11, and one uranyl coprecipitation experiment, U1, to enable comparison of neptunyl and uranyl coprecipitation.

In general all MFR experiments are carried out under a set of standard conditions. Only single parameters are modified to study their influence. Temperature in the water bath is 25 ± 0.1°C during all the experiments. Stirring velocity in the MFR is ~ 850 rpm. Flowrate through the MFR is adjusted to an initial flowrate, F, of about 0.6 mL/min. This corresponds to a mean residence time of the solution in the reactor of 75 min. 160 mg crystal seeds are introduced into the MFR before the experiments. This corresponds to a reactive surface area of about 0.2 m^2 in the MFR. 0.01 M NaCl are added to all three stock

3. Experimental details

solutions as background electrolyte. Solution 1 contains additionally $2.4 \cdot 10^{-3}$ M Na_2CO_3 (designated below as total inorganic carbon (TIC)), solution 2 contains the dopant, and solution 3 contains $9 \cdot 10^{-4}$ M $CaCl_2$. Mixing of these three solutions defines the standard input solution conditions:

- $c(NaCl) = 0.01$ M
- $c(Ca^{2+}) = 0.0003$ M
- $c(TIC) = 0.0008$ M
- pH ≈ 10.4
- ratio $\frac{HCO_3^-}{CO_3^{2-}} \approx 0.6$
- SI(calcite) ≈ 1.1
- Ionic strength ≈ 0.013 M

In the reactor, calcite precipitates and the Ca^{2+} and TIC concentrations decrease by the same amount. Accordingly pH and SI(calcite) decrease, too. Steady state SI and steady state growth rate are the result of a complicated interplay between SI of the input solution, flowrate, and growth retardation by the dopant. They are reproducible but cannot be predicted.

Cc1 - Cc4 : During the four experiments with pure calcite all experimental parameters are kept constant except the flowrate. Flowrate is varied from 0.4 to 1.6 mL/min. This causes the steady state SI from 0.04 to 0.53 and growth rate to vary from $1.2 \cdot 10^{-8}$ mol/(m²s) to $6.6 \cdot 10^{-8}$ mol/(m²s).

Np1 - Np4 : These experiments have been conducted at 1 μM neptunyl concentration in the MFR (3μM in the stock solution). Through this series of experiments, the pH has been varied from 8.2 to 12.8. Experiments **Np1** and **Np2** have been conducted at the standard conditions described above plus 1 μM neptunyl. In experiment **Np3**, the pH has been adjusted to 12.8 by modifying the carbonate solution to a composition of 0.01 M NaCl, 0.003 M Na_2CO_3, and 0.3 M NaOH. To achieve an input SI close to one, the calcium concentration has been increased to 0.003 M. In experiment **Np4** the pH has been adjusted to 8.2 by modifying the carbonate solution to a composition

3. Experimental details

of 0.01 M NaCl and 0.0024 M NaHCO$_3$. To achieve an input SI close to one, the calcium concentration has been increased to 0.006 M.

U1 : For the uranyl coprecipitation experiment the same experimental conditions have been chosen as in experiment Np5 except that 1 μM uranyl has been used instead of neptunyl.

Np5 - Np10 : This series of experiments has been used to check the influence of varying aqueous neptunyl / calcium concentration ratios. This has been actualized by changing the neptunyl input concentration from $1 \cdot 10^{-7}$ M ($3 \cdot 10^{-7}$ M in the stock solution) in experiments Np5 to Np7 to $5 \cdot 10^{-6}$ M ($1.5 \cdot 10^{-5}$ M in the stock solution) in experiments Np8 to Np10.

Np11(SC) : Experiment Np11 has been used to synthesize calcite single crystals covered with a thin layer of neptunyl doped calcite. In order to adjust the growth conditions to the reduced reactive surface area in the MFR ($\sim$ 0.002 m^2) the SI of the input solution has been reduced to 0.4 and flowrate has been reduced to 0.01 mL/min.

The most important experimental parameters are summarized in Table 3.2.

3.9.2. EXAFS characterization of calcites from experiments Np1 – Np4 and U1

These EXAFS measurements have been performed at the INE-Beamline for actinide research at ANKA [107] in fluorescence-mode using a solid state detector (LEGe Canberra). For measurements at the Np L$_3$ edge (17.610 keV), we use Ge(422) monochromator crystals. Energy calibration is done by parallel measurement of a Zr-foil, defining the first inflection point in the Zr K-edge as 17.998 keV. The uranium L$_3$-EXAFS (L$_3$ edge at 17.166 keV) is measured using Si(311) monochromator crystals, and a Y-foil for energy calibration (Y K-edge, 17.038 keV). Seven to seventeen spectra have been averaged and analyzed using backscattering amplitude and phase shift functions calculated with Feff 8 [108] and the FEFFIT 2.54 software. Background removal is done with WinXAS 3.1 [104]. The maximum of the most intense absorption feature, the so called white line, is set to the ionization energy, E_0, which serves as the origin for generating k-values. In the fit procedure E_0 is modified by addition of the relative shift in ionization energy, ΔE_0. The amplitude reduction factor, S_0^2, is adjusted to 0.8 for Np and 0.9 for U, values obtained in preliminary

3. Experimental details

Tab. 3.2: Summary of experimental conditions for the MFR experiments.

Experiment	c_{in}(TIC) (mmol/L)	c_{in}(Ca) (mmol/L)	c_{in}(Np/U) (μmol/L)	pH_{in}	SI_{in}	F (mL/min)
Cc1	0.8	0.28	0	10.4	1.1	0.61
Cc2	0.8	0.27	0	10.4	1.1	0.44
Cc3	0.8	0.27	0	10.4	1.1	1.30
Cc4	0.8	0.27	0	10.4	1.1	1.63
Np1	0.8	0.34	0.92	10.4	1.1	0.52
Np2	0.8	0.35	0.80	10.4	1.2	0.50
Np3	1.0	1.30	0.87	12.8	1.3	0.60
Np4	8.0	2.36	1.10	8.2	1.3	0.53
Np5	0.8	0.30	0.12	10.4	1.1	0.56
Np6	0.8	0.30	0.12	10.4	1.1	0.55
Np7	0.8	0.30	0.12	10.4	1.1	0.37
Np8	0.8	0.31	5.08	10.4	1.1	0.55
Np9	0.8	0.31	5.08	10.4	1.1	0.54
Np10	0.8	0.31	5.08	10.4	1.1	0.37
Np11(SC)	0.3	0.16	1.10	10.3	0.4	0.01
U1	8.0	2.80	1.01	8.2	1.4	0.58

fits assuming that the number of the axial oxygens is two. A S_0^2 of 0.9 for uranium is in good agreement with published theoretical and experimental values [109, 110]. Fits are performed in R-space in the range between 0.7 and 3.5 Å. A Hanning window function is used for the Fourier transformation in a k-range of 2.4 to $\sim$12 Å^{-1} for the neptunium spectra and 1.8 to 8.5 Å^{-1} for the uranium spectrum. Only single scattering paths are used to fit the data. A global relative shift in ionization energy, ΔE_0, used is 9.2 eV for the neptunium samples and 8.42 eV for the uranium sample. These values are also obtained in preliminary fits to the data. For the neptunyl results bond-valence calculations [96] have been used to check the plausibility of the obtained coordination environment around the central neptunium atom.

3.9.3. Near infrared (NIR) spectroscopy

For the f^2 electronic configuration of neptunyl(V) (see Figure 1.1) all of the excited states in the infrared and visible range contain two f-electrons, thus f→f transitions to these states are in principle electric dipole forbidden by Laporte's rule. However, this rule is followed only strictly if there is inversion symmetry in the coordination environment around the neptunium ion. In the NIR region aqueous neptunyl(V) has a characteristic absorption band at 980 nm that is usually shifted to higher wavelengths when water in the first coordination sphere is replaced by coordinating ligands [111, 112].

NIR absorption spectra of the single crystals synthesized in MFR experiment Np11 have been recorded on a Cary 5 UV-Vis-NIR spectral photometer. Background subtraction is performed by parallel measurements of pure calcite crystals of similar dimensions in the second beam of the photometer.

3.9.4. Raman spectroscopy

The ν_1 symmetric stretching vibrational mode of the neptunyl(V) ion shows a chemical shift with changes in the equatorial coordination environment. Due to the f^2 electronic configuration of neptunyl(V) ion this shift is small compared to uranyl(VI) (f^0) [112]. However, the neptunyl ν_1 vibrational mode does undergo a measurable chemical shift upon carbonate coordination from 767 cm^{-1} to 756 cm^{-1} that is inversely correlated to the distance between neptunium and the axial neptunyl oxygen atoms (1.82 Å– 1.86 Å) [113]. For comparison, the corresponding shift for uranyl(VI) is from 870 cm^{-1} to 812 cm^{-1} [112]. Raman spectra of pure calcite and neptunyl doped calcite from MFR experiments Np8 - Np10 have been collected on a Bruker Senterra Raman Microscope using green diode laser light with a wavelength of 532 nm.

Chapter 4

Results and discussion

4.1. The calcite surface

4.1.1. Zetapotential

The results of the electrophoretic PALS zetapotential measurements performed on calcite particles in equilibrium with solution and a gas phase are shown in Figure 4.1. In each of the three datasets the zetapotential decreases from positive values around 12 mV at the lowest pH values to negative values around -6 mV at the highest pH values. With decreasing CO_2 partial pressure the zetapotential curve shifts towards higher pH. In equilibrium with CO_2 (red dots in Figure 4.1) the isoelectric point (IEP) where the zetapotential is 0 mV is at about pH 6.5. In equilibrium with air (green dots in Figure 4.1) IEP is at about pH 8.8. In equilibrium with N_2 (blue dots in Figure 4.1) IEP is at about pH 9.4. This corresponds to a nearly linear dependence between $\log_{10}(p(CO_2))$ and IEP.

A completely different behavior can be observed when measuring zetapotential in non–equilibrium solutions as shown in Figure 4.2 for the results of the streaming potential measurements. Titrations in 0.01 M NaCl solution (blue dots in Figure 4.2) show that the zetapotential is nearly constant throughout the whole pH region investigated. In the neutral pH region a plateau value of about -16 mV, above pH 9 a slight decrease to minimal values of about -20 mV at pH 11, and below pH 6 a slight increase are observed. The purple data points correspond to a titration with 5 mM Na_2CO_3 in solution. At low pH the values are to within experimental uncertainty the same as measurements without carbonate. Above pH 8, where deprotonation of bicarbonate begins, the zetapotential

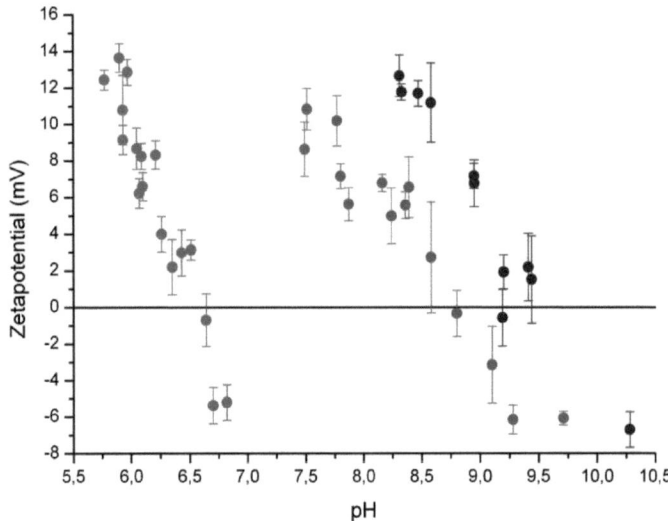

Fig. 4.1: Results of electrophoretic PALS zetapotential measurements in equilibrium solutions at an ionic strength of 0.1 M NaCl at various solution compositions and CO_2 partial pressures. Red: equilibrium with CO_2 ($p(CO_2) = 1$ bar), green: equilibrium with air (360 ppm CO_2), and blue: equilibrium with N_2 (6 ppm CO_2).

starts to decrease significantly to values of about -25 mV at pH 11. The opposite trend is observed for subsequent titrations with increasing $CaCl_2$ concentration in solution (green, yellow, orange, and red data points in Figure 4.2). A continuous increase in zetapotential is observed suggesting that carbonate and calcium adsorb at the surface and cause changes in zetapotential, while protonation and deprotonation reactions have only a limited influence on the calcite zetapotential. As already discussed in section 2.2.2 similar results have been previously reported [64, 65]. Unexpected behavior occurs at elevated calcium concentration. Depending on the direction and the starting pH of of the titrations, a hysteresis effect is observed. Open symbols in Figure 4.2 mark titrations starting at high pH that show hysteresis compared to the titrations starting at low pH. In titrations starting at low

4. Results and discussion

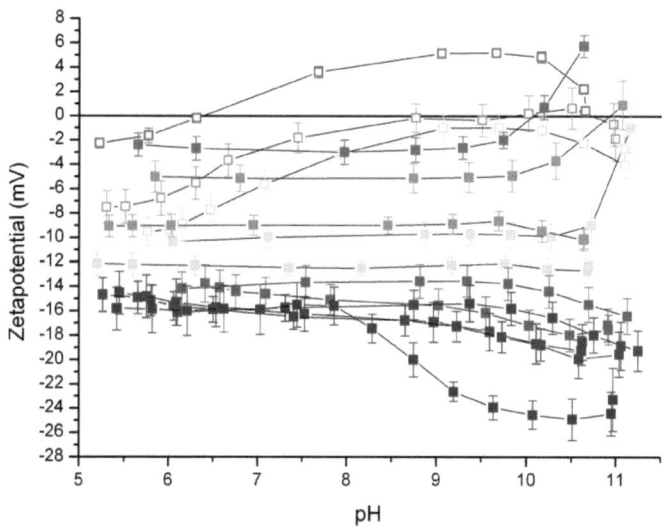

Fig. 4.2: Streaming potential measurements on calcite in non equilibrium solutions. Color coded datasets correspond to titrations with HCl or NaOH. Background electrolyte concentration is 0.01 M NaCl. Purple: 5 mM Na_2CO_3, blue: no additional ions in solution, green: 0.1 mM $CaCl_2$, yellow: 0.5 mM $CaCl_2$, orange: 1 mM $CaCl_2$, and red: 5 mM $CaCl_2$.

pH the zetapotential remains constant over a large pH range and then shows a sudden increase at a certain pH value, pH_c, the value of which depends on calcium concentration. At $c(Ca^{2+}) = 0.5$ mM $pH_c \sim 10.9$, at $c(Ca^{2+}) = 1$ mM $pH_c \sim 10.7$, at $c(Ca^{2+}) = 5$ mM $pH_c \sim 10$. In titrations in the opposite direction, i.e. starting above pH_c, the zetapotential begins and remains at high values until a pH of about 9 where it slowly decreases (open symbols in Figure 4.2). Curves of forward and backward titrations intersect between pH 5.5 and 6. If the starting pH for a downwards titration is below pH_c (examples are shown in the yellow and orange datasets) a completely flat zetapotential curve is observed.

Similar effects have been observed recently on fused silica surfaces [114]. There it has been shown that, depending on the electrolyte concentration, changes in surface speciation

4. Results and discussion

induced by pH changes occur with a time delay of up to more than one hour.

Formation of calcium hydroxide surface precipitates has been another hypothesis to explain the observed changes in zetapotential at high pH and high calcium concentrations, even though the calcium and hydroxide concentrations are still far below the portlandite solubility limit. A surface precipitate with a 1:1 ratio of Ca^{2+} and OH^- would be highly positive charged and could explain the change in zetapotential. Assuming a slow dissolution kinetics of such a precipitate could also explain the hysteresis effect.

These requests underpin the great care that must be taken when considering experimental data on surface charge and potential for surface complexation modeling that have been recorded in non–equilibrium solutions. The streaming potential measurements show that the changes in solution composition due to calcite dissolution are neglible (cp. section 3.5.2). Nevertheless, effects like time delay of surface protonation / deprotonation reactions or slow dissolution kinetics of a surface precipitate could adulterate the results.

To get a better understanding of the processes at the calcite surface, structural changes at the calcite–water interface potentially induced by calcium and carbonate adsorption have been explored by a surface diffraction study. Inner–sphere adsorption and especially surface precipitation, which might take place at high pH and high calcium concentration, can be clearly detected with this method.

4.1.2. Results from surface diffraction

The 3D structure of the calcite–water interface has been investigated by means of surface diffraction measurements at six selected solution conditions. The solution conditions are summarized in Table 4.1. pH measurements indicate if the solutions are well equilibrated. Measured values are compared to the expected values given in section 3.6.3. Only solution G showed a significantly higher pH than expected. Solution G in equilibrium with CO_2 (1 bar) probably degassed during the sample transport. The pH value expected at 1 bar CO_2 is 5.8. The observed pH, 6.3, would be expected for a CO_2 partial pressure of 0.1 bar. Therefore this partial pressure is assumed as equilibrium pressure in the speciation calculations in Table 4.1.

The solution speciation concerning calcium concentration and the concentration of the different carbonate species is shown in Table 4.1 together with pH, ionic strength, and the zetapotential expected based on the measurements described above (see section 4.1.1).

4. Results and discussion

Tab. 4.1: Solution– and surface speciation (after the Pokrovsky and Schott SCM [62]) at the conditions chosen for surface diffraction measurements.

	B	C	D	E	F	G
pH	8.6	7.5	10.9	11.1	8.6	6.3
IS (mol/L)	0.1	0.1	0.04	0.04	0.002	0.1
zetapotential (mV)	~ 0	~ 9	~ 8	~ -25	~ 0	~ 10
$c(Ca^{2+})$ (mmol/L)	0.21	34.99	10.00	0.00	0.12	36.88
$c(CO_3^{2-})$ (mmol/L)	0.13	< 0.01	0.00	9.06	0.04	< 0.01
$c(HCO_3^-)$ (mmol/L)	2.98	0.25	0.00	0.94	2.10	4.04
$F(>CO_3H)$ (%)	< 1	< 1	< 1	< 1	< 1	< 1
$F(>CO_3^-)$ (%)	67	14	26	100	52	13
$F(>CO_3Ca^+)$ (%)	33	86	74	< 1	48	86
$F(>CaOH)$ (%)	< 1	< 1	18	< 1	< 1	< 1
$F(>CaO^-)$ (%)	< 1	< 1	3	< 1	< 1	< 1
$F(>CaOH_2^+)$ (%)	25	80	79	1	38	73
$F(>CaHCO_3^0)$ (%)	< 1	< 1	0	< 1	< 1	< 1
$F(>CaCO_3^-)$ (%)	75	20	0	99	61	17

The solutions where zetapotential is high (C, D, and G) are characterized by high calcium concentrations. In the equilibrium solutions B, C, F, and G the carbonate speciation is dominated by bicarbonate. Solution E is the only one for which a negative zetapotential is expected, due to the absence of calcium and the high carbonate concentration.

The molar fractions of surface complexes expected due to the Pokrovsky and Schott SCM [62] are listed below the solution speciation in Table 4.1. In solutions C, D, and G about 80 % of the carbonate surface groups are expected to be complexed with calcium. In solution E 99 % of the surface calcium atoms are complexed with carbonate ($>CaCO_3^-$) and the surface carbonate groups are predicted to be 100 % $>CO_3^-$. This means nearly all of the surface sites are negatively charged. This shows how extreme the chosen conditions in solutions C, D, and G are compared to E with respect to the expected surface speciation.

4. Results and discussion

Such high coverages with inner–sphere surface complexes would have a significant impact on surface diffraction data. Fenter et al. (2000) [48] show the influence of inner–sphere complexes on the specular calcite CTR and estimate the uncertainty for the determination of the site density of inner–sphere copmplexes from surface diffraction data to be about 4 % – 9 %. This range of uncertainty values is in good agreement with observations during structure refinement in this study.

All the extracted data of surface diffraction measurements are shown in Figure 4.3. It is

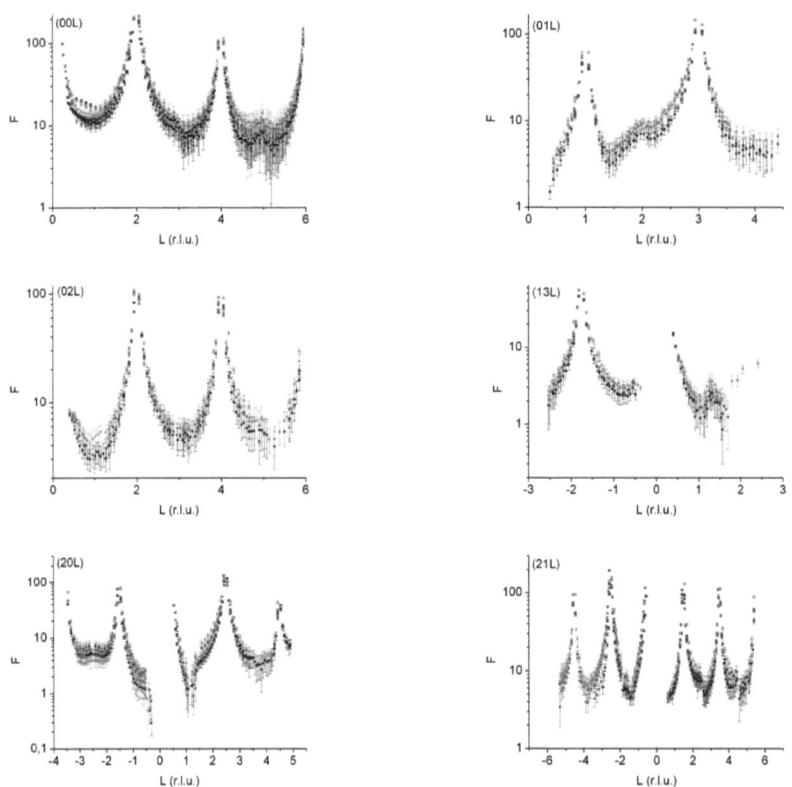

Fig. 4.3: Overview over CTR data of all datasets. B: dark green, C: orange, D: blue, E: purple, F: green, G: red

4. Results and discussion

surprising that the CTRs measured on calcite in the solutions described above look nearly identical. In fact a very detailed look at the data is required to discover any significant differences between the different datasets. As each dataset has its own scaling factor, the relative shifts in structure factor magnitude are not related to structural differences. Only variations in the shapes of the CTRs are related to variations in the structure. In the low L region between L = 0.5 and L = 1.5 on the (00L) rod the differences between the datasets are most likely to be significant. The deviation in the low L region on the (02L) rod in dataset C is probably related to contributions of a powder line to the data. Another example for differences that might be significant can be found around L = -3 on the (21L) rod.

Preliminary fitting approaches have shown that all the datasets are indeed very similar and they are all modeled best without considering any inner–sphere sorbed calcium or carbonate ions. The surface model which enabled reasonable fitting of all the CTR datasets included two relaxable calcite monolayers. One directly at the surface (1st calcite layer) and one below (2nd calcite layer) and two well ordered layers of water above the surface. $H_2O(1)$ above surface calcium atoms and $H_2O(2)$ above surface carbonate atoms. In order to obtain good fits for the low L region on the (00L) rod (L < 1.5), two more layers of water have to be included into the model. These layers are characterized by very high Debye–Waller factors. Water molecules in these layers cannot be regarded as structured. They more likely resemble liquid water and show the influence of diffuse electron density on surface diffraction data. Modeling this part of the structure by two additional water molecules is surely an oversimplification, but the high Debye–Waller factors make fitting of more complicated and reasonable structures impossible.

According to the fit file (see appendix A) the top two layers of the calcite bulk structure are treated as adjustable. Water molecules of the first water layer, depicted by oxygen atoms in the fit model, are initially located where the oxygen atoms of the next carbonate ions in the calcite bulk structure next to the surface calcium atoms would be expected. This results in an octahedral coordination of the surface calcium atom. Water molecules of the second water layer are located 2.3 Å above the outwards oriented oxygen atoms of the surface carbonate groups. Initial location of the third and fourth water layer is in a height of 4.9 Å above the surface (defined by the z–position of the surface Ca atoms) above the water molecules of the first and second water layer.

The structural parameters are adjusted to give an optimal fit between calculated and mea-

4. Results and discussion

sured CTRs. CTRs of one dataset are modeled simultaneously. Due to the similarity between the datasets, the best modeling approach turned out to be, to first fit an average structure refined on all the datasets together. The structure obtained in this fit is then used as starting point for subsequent fitting of the individual datasets. Using this approach differences between the resulting structures caused by correlated parameters can be minimized.

4. Results and discussion

Solution B

CTRs measured on calcite in solution B and the corresponding modeling results are shown in Figure 4.4. The roughness value β obtained for this dataset is: $\beta = 0.03 \pm 0.01$. This corresponds to a root mean square roughness, σ_{rms} of: $\sigma_{rms} = 1.0 \pm 0.5$ Å. The goodness of fit measured as normalized χ^2 is: $\chi^2 = 5.33$. The bond valence sum for calcium in the top calcite monolayer is: $BVS(Ca) = 2.00$. The structural fitting parameters are listed in Table 4.2. Dataset B contains 486 data points.

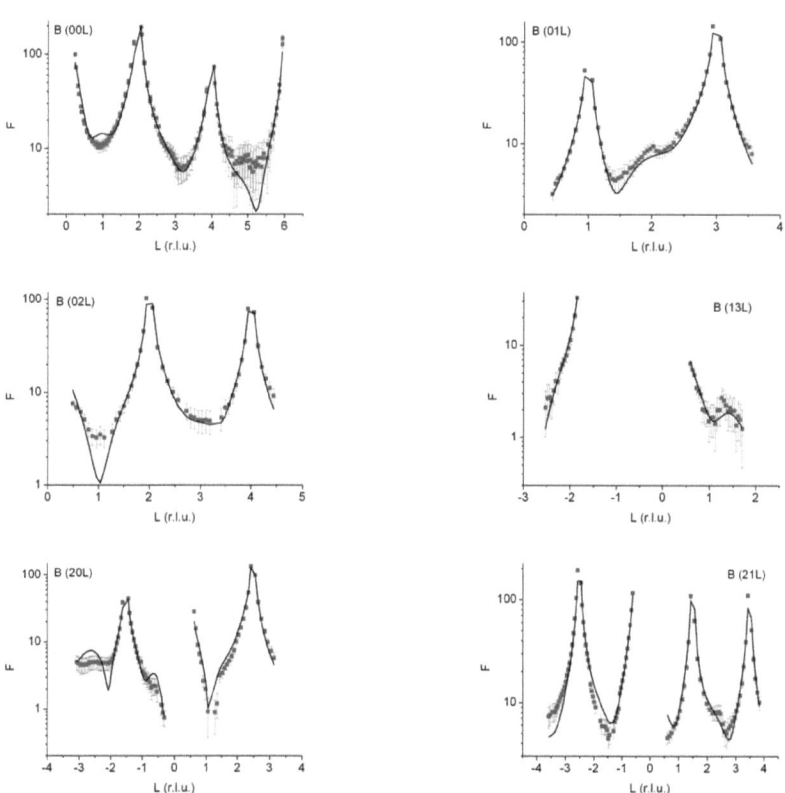

Fig. 4.4: CTRs of dataset B, data (dark green squares) and model (black line).

4. Results and discussion

The molecular surface structure according to the parameters listed in Table 4.2 is illustrated in Figure 4.5.

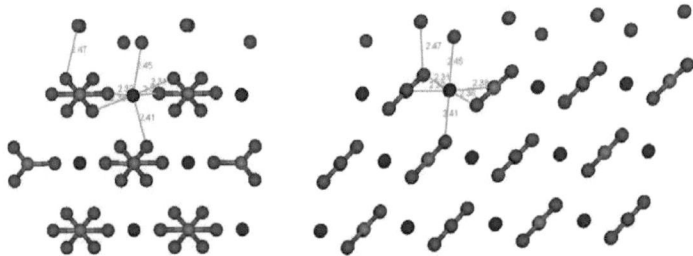

Fig. 4.5: Molecular structure of the calcite–water interface according to the result of the structure refinement of dataset B. The left hand side shows a projection along the b_1 direction, the right hand side shows a projection along the b_2 direction. Selected interatomic distances are indicated. (blue: Ca, gray: C, red: O)

Tab. 4.2: Structural parameters obtained from fitting of dataset B. Uncertainties are twice the standard deviation calculated by the ROD software. Displacement parameters Δx, Δy, and Δz are in fractional coordinates, B_j are Debye–Waller factors in Å^2. θ and ϕ are the carbonate rotation angles in degrees (see Figure 3.6 for definition).

	Δx (b_1)	Δy (b_2)	Δz (b_3)	B_j
	structured water			
H$_2$O(2)	0.015 ± 0.003	-0.12 ± 0.02	0.023 ± 0.004	12 ± 2
H$_2$O(1)	0.011 ± 0.005	-0.21 ± 0.02	0.024 ± 0.005	12 ± 1
	1st calcite layer			
Ca	0.008 ± 0.001	-0.002 ± 0.002	0.006 ± 0.001	2.4 ± 0.3
CO$_3$	0.014 ± 0.003	-0.005 ± 0.004	0.011 ± 0.003	6.2 ± 0.5
	$\theta = -3.5 \pm 1.1$,	$\phi = -1.5 \pm 0.8$		
	2nd calcite layer			
Ca	0.004 ± 0.001	-0.001 ± 0.001	0.002 ± 0.001	0.5 ± 0.2
CO$_3$	0.002 ± 0.002	-0.008 ± 0.003	-0.003 ± 0.002	0.9 ± 0.3
	$\theta = -1.5 \pm 0.7$,	$\phi = -0.9 \pm 0.5$		

4. Results and discussion

Solution C

CTRs measured on calcite in solution C and the corresponding modeling results are shown in Figure 4.6. Dataset C contains 469 data points.

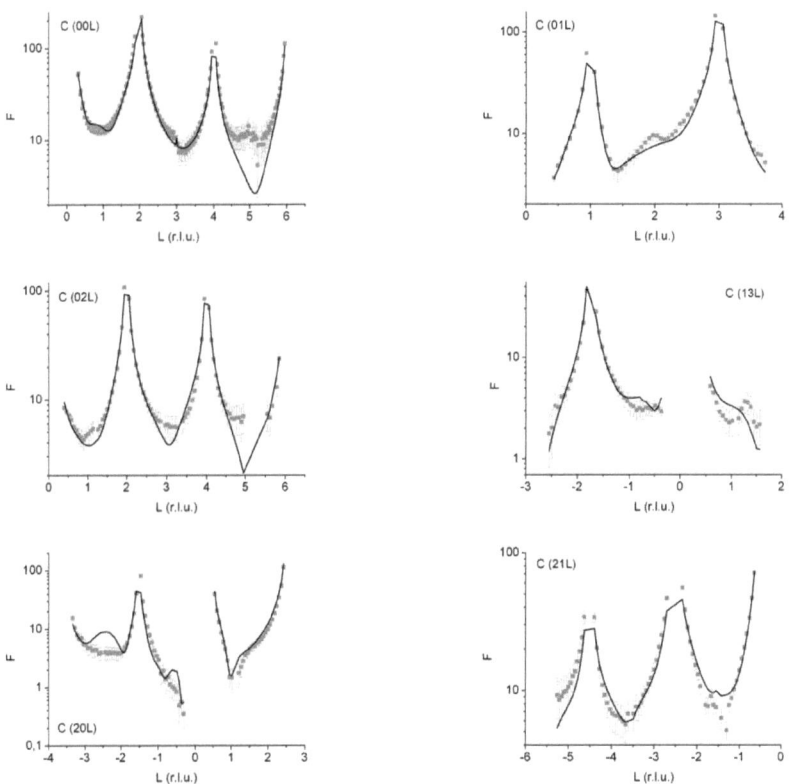

Fig. 4.6: CTRs of dataset C, data (orange squares) and model (black line).

The roughness value obtained for this dataset is: $\beta = 0.14 \pm 0.01$. This corresponds to a root mean square roughness: $\sigma_{rms} = 2.6 \pm 0.6$ Å. The goodness of fit is: $\chi^2 = 3.89$. The bond valence sum for calcium in the top calcite monolayer is: BVS(Ca) = 2.18. The structural fitting parameters are listed in Table 4.3.

4. Results and discussion

The molecular surface structure according to the parameters listed in Table 4.3 is illustrated in Figure 4.7.

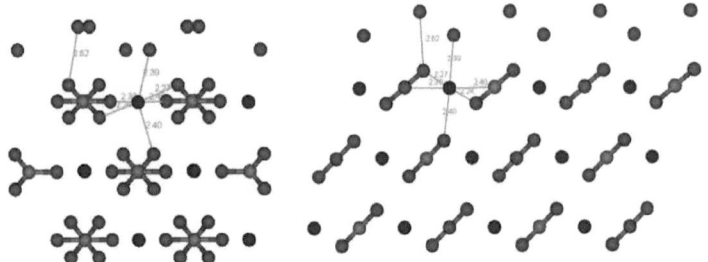

Fig. 4.7: Molecular structure of the calcite–water interface according to the result of the structure refinement of dataset C. The left hand side shows a projection along the b_1 direction, the right hand side shows a projection along the b_2 direction. Selected interatomic distances are indicated. (blue: Ca, gray: C, red: O)

Tab. 4.3: Structural parameters obtained from fitting of dataset C. Uncertainties are twice the standard deviation calculated by the ROD software. Displacement parameters Δx, Δy, and Δz are in fractional coordinates, B_j are Debye–Waller factors in Å^2. θ and ϕ are the carbonate rotation angles in degrees (see Figure 3.6 for definition).

	Δx (b_1)	Δy (b_2)	Δz (b_3)	B_j
	structured water			
$H_2O(2)$	-0.011 ± 0.006	-0.07 ± 0.02	0.018 ± 0.006	20 ± 2
$H_2O(1)$	-0.010 ± 0.005	-0.238 ± 0.008	0.050 ± 0.008	8.7 ± 0.7
	1st calcite layer			
Ca	-0.002 ± 0.001	-0.004 ± 0.001	0.002 ± 0.001	2.0 ± 0.2
CO_3	0.004 ± 0.003	0.011 ± 0.002	-0.003 ± 0.002	6.3 ± 0.4
	$\theta = -4.7 \pm 1.2$,	$\phi = 0.2 \pm 0.8$		
	2nd calcite layer			
Ca	-0.003 ± 0.002	0.000 ± 0.001	0.000 ± 0.001	0.1 ± 0.2
CO_3	-0.002 ± 0.002	0.013 ± 0.004	0.007 ± 0.003	1.0 ± 0.3
	$\theta = -0.7 \pm 0.7$,	$\phi = 0.1 \pm 0.5$		

4. Results and discussion

Solution D

CTRs measured on calcite in solution D and the corresponding modeling results are shown in Figure 4.8. Dataset D contains 590 data points; it is the largest dataset.

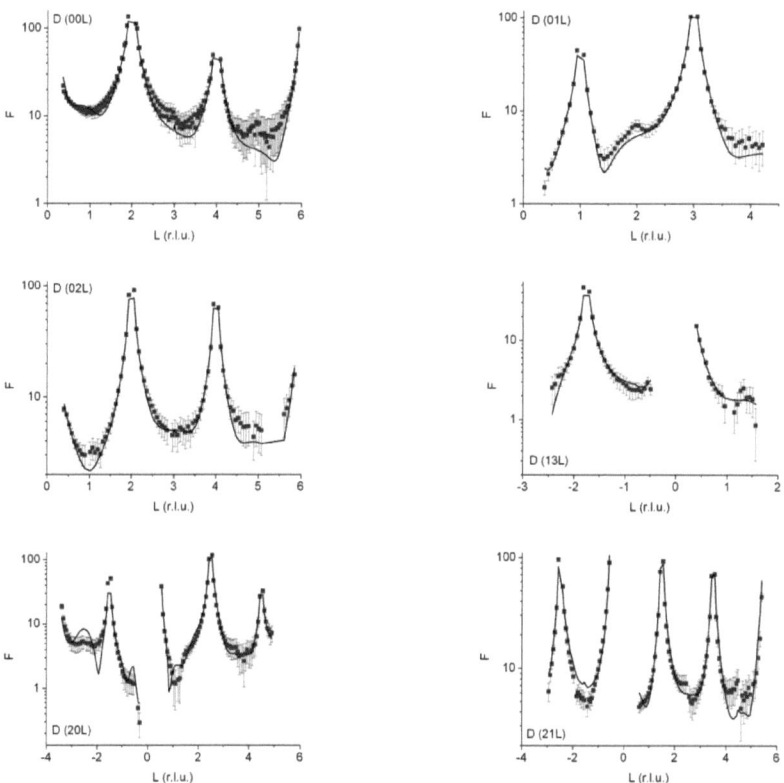

Fig. 4.8: CTRs of dataset D, data (blue squares) and model (black line).

The roughness value obtained for this dataset was: $\beta = 0.23 \pm 0.01$. This corresponds to a root mean square roughness: $\sigma_{rms} = 3.8 \pm 0.6$ Å. The goodness of fit is: $\chi^2 = 3.36$, and the bond valence sum for calcium in the top calcite monolayer was: BVS(Ca) = 2.17. The structural fitting parameters are listed in Table 4.4.

4. Results and discussion

The molecular surface structure according to the parameters listed in Table 4.4 is illustrated in Figure 4.9.

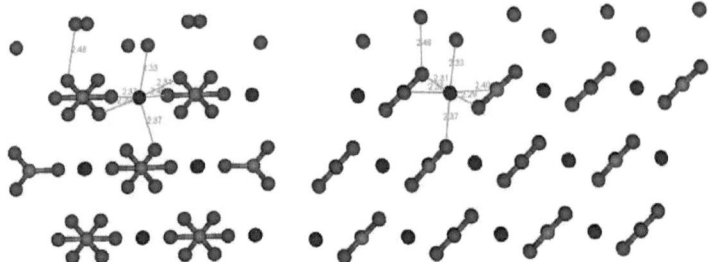

Fig. 4.9: Molecular structure of the calcite–water interface according to the result of the structure refinement of dataset D. The left hand side shows a projection along the b_1 direction, the right hand side shows a projection along the b_2 direction. Selected interatomic distances are indicated. (blue: Ca, gray: C, red: O)

Tab. 4.4: Structural parameters obtained from fitting of dataset D. Uncertainties are twice the standard deviation calculated by the ROD software. Displacement parameters Δx, Δy, and Δz are in fractional coordinates, B_j are Debye–Waller factors in Å^2. θ and ϕ are the carbonate rotation angles in degrees (see Figure 3.6 for definition).

	Δx (b_1)	Δy (b_2)	Δz (b_3)	B_j
	structured water			
$H_2O(2)$	-0.003 ± 0.006	-0.08 ± 0.03	0.001 ± 0.007	19 ± 2
$H_2O(1)$	-0.003 ± 0.000	-0.22 ± 0.01	0.035 ± 0.006	9.1 ± 0.1
	1st calcite layer			
Ca	-0.011 ± 0.002	-0.001 ± 0.001	-0.001 ± 0.001	1.2 ± 0.3
CO_3	-0.008 ± 0.002	0.009 ± 0.004	0.011 ± 0.003	5.9 ± 0.5
	$\theta = -2.8 \pm 1.5$,		$\phi = 0.2 \pm 1.0$	
	2nd calcite layer			
Ca	-0.006 ± 0.003	-0.004 ± 0.002	-0.003 ± 0.003	1.3 ± 0.2
CO_3	-0.006 ± 0.001	-0.002 ± 0.003	0.001 ± 0.001	1.2 ± 0.3
	$\theta = -0.8 \pm 0.7$,		$\phi = 0.6 \pm 0.9$	

4. Results and discussion

Solution E

CTRs measured on calcite in solution E and the corresponding modeling results are shown in Figure 4.10. Dataset E contains 283 data points.

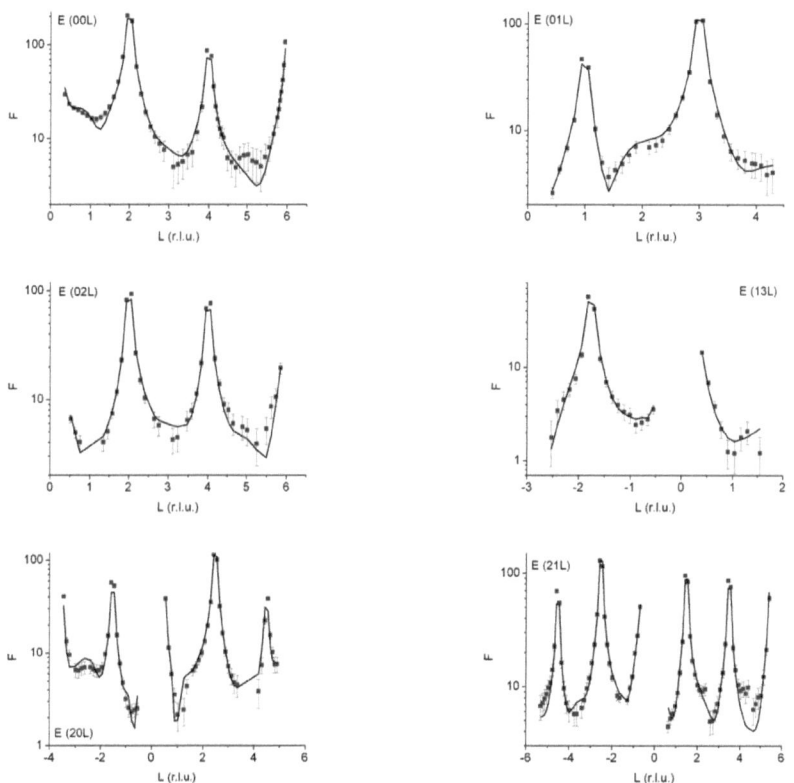

Fig. 4.10: CTRs of dataset E, data (purple squares) and model (black line).

The roughness value obtained for this dataset was: $\beta = 0.07 \pm 0.01$. This corresponds to a root mean square roughness: $\sigma_{rms} = 1.7 \pm 0.6$ Å. The goodness of fit was: $\chi^2 = 3.91$, and the bond valence sum for calcium in the top calcite monolayer was: BVS(Ca) = 2.09. The structural fitting parameters are listed in Table 4.5.

4. Results and discussion

The molecular surface structure according to the parameters listed in Table 4.5 is illustrated in Figure 4.11.

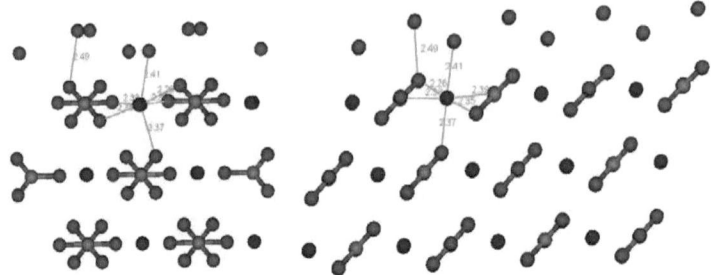

Fig. 4.11: Molecular structure of the calcite–water interface according to the result of the structure refinement of dataset E. The left hand side shows a projection along the b_1 direction, the right hand side shows a projection along the b_2 direction. Selected interatomic distances are indicated. (blue: Ca, gray: C, red: O)

Tab. 4.5: Structural parameters obtained from fitting of dataset E. Uncertainties are twice the standard deviation calculated by the ROD software. Displacement parameters Δx, Δy, and Δz are in fractional coordinates, B_j are Debye–Waller factors in Å^2. θ and ϕ are the carbonate rotation angles in degrees (see Figure 3.6 for definition).

	Δx (b_1)	Δy (b_2)	Δz (b_3)	B_j
	structured water			
$H_2O(2)$	-0.006 ± 0.006	-0.07 ± 0.02	0.033 ± 0.006	13 ± 2
$H_2O(1)$	-0.022 ± 0.006	-0.217 ± 0.014	0.008 ± 0.008	12 ± 1
	1st calcite layer			
Ca	0.004 ± 0.002	-0.008 ± 0.001	-0.005 ± 0.001	1.6 ± 0.2
CO_3	0.008 ± 0.003	0.007 ± 0.004	0.010 ± 0.003	5.4 ± 0.2
	$\theta = -2.6 \pm 1.6$,		$\phi = -1.9 \pm 1.0$	
	2nd calcite layer			
Ca	0.003 ± 0.001	-0.004 ± 0.001	0.001 ± 0.001	0.8 ± 0.2
CO_3	-0.005 ± 0.002	0.001 ± 0.004	-0.005 ± 0.003	0.8 ± 0.3
	$\theta = -0.8 \pm 0.7$,		$\phi = 0.8 \pm 0.9$	

4. Results and discussion

Solution F

CTRs measured on calcite in solution F and the corresponding modeling results are shown in Figure 4.12. Dataset F contains 278 data points.

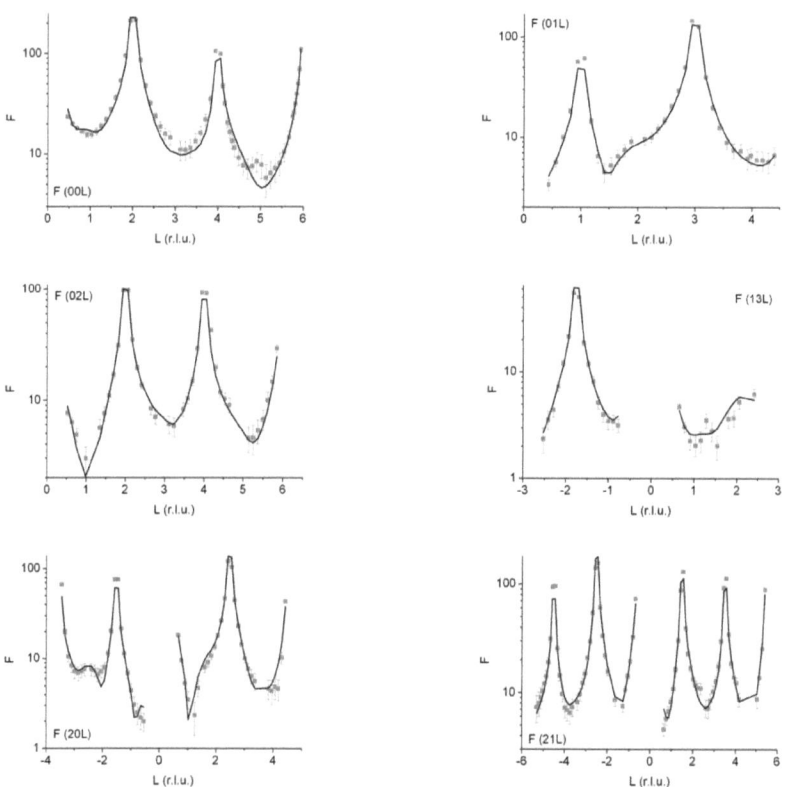

Fig. 4.12: CTRs of dataset F, data (green squares) and model (black line).

The roughness value obtained for this dataset was: $\beta = 0.01 \pm 0.01$. This corresponds to a root mean square roughness: $\sigma_{rms} = 0.6 \pm 0.5$ Å. The goodness of fit was: $\chi^2 = 7.75$, and the bond valence sum for calcium in the top calcite monolayer was: BVS(Ca) = 2.14. The structural fitting parameters are listed in Table 4.6.

4. Results and discussion

The molecular surface structure according to the parameters listed in Table 4.6 is illustrated in Figure 4.13.

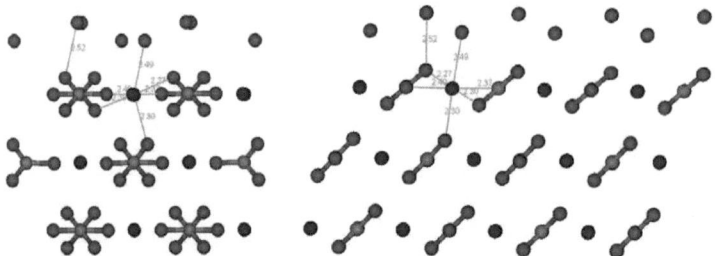

Fig. 4.13: Molecular structure of the calcite–water interface according to the result of the structure refinement of dataset F. The left hand side shows a projection along the b_1 direction, the right hand side shows a projection along the b_2 direction. Selected interatomic distances are indicated. (blue: Ca, gray: C, red: O)

Tab. 4.6: Structural parameters obtained from fitting of dataset F. Uncertainties are twice the standard deviation calculated by the ROD software. Displacement parameters Δx, Δy, and Δz are in fractional coordinates, B_j are Debye–Waller factors in Å^2. θ and ϕ are the carbonate rotation angles in degrees (see Figure 3.6 for definition).

	Δx (b_1)	Δy (b_2)	Δz (b_3)	B_j
	structured water			
$H_2O(2)$	0.003 ± 0.004	-0.11 ± 0.01	0.024 ± 0.004	10 ± 1
$H_2O(1)$	-0.012 ± 0.006	-0.233 ± 0.011	0.020 ± 0.007	10 ± 1
	1st calcite layer			
Ca	-0.004 ± 0.002	-0.007 ± 0.001	-0.005 ± 0.001	1.1 ± 0.2
CO_3	-0.004 ± 0.002	0.010 ± 0.003	0.000 ± 0.000	5.3 ± 0.5
	$\theta = -3.8 \pm 1.3$,	$\phi = 0.2 \pm 0.8$		
	2nd calcite layer			
Ca	-0.002 ± 0.001	-0.003 ± 0.001	0.000 ± 0.001	0.1 ± 0.2
CO_3	-0.004 ± 0.002	-0.001 ± 0.003	-0.003 ± 0.002	0.7 ± 0.3
	$\theta = 1.1 \pm 0.8$,	$\phi = 1.4 \pm 0.6$		

Solution G

CTRs measured on calcite in solution G and the corresponding modeling results are shown in Figure 4.14. Dataset G is with 230 data points the smallest dataset. The (13L) rod was not measured on this sample.

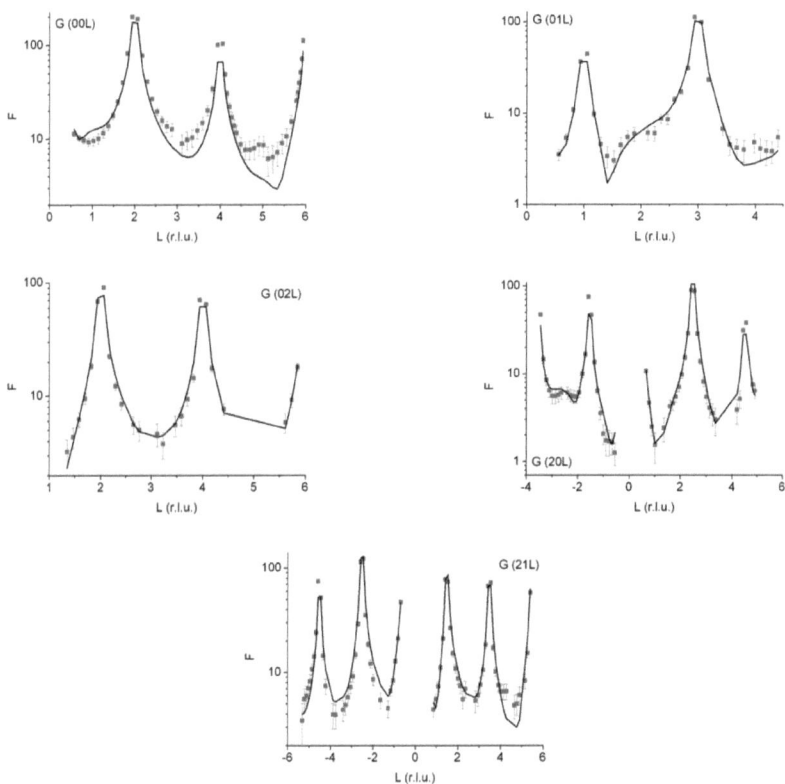

Fig. 4.14: CTRs of dataset G, data (red squares) and model (black line).

The roughness value obtained for this dataset was: $\beta = 0.04 \pm 0.01$. This corresponds to a root mean square roughness: $\sigma_{rms} = 1.2 \pm 0.6$ Å. The goodness of fit was: $\chi^2 = 10.4$, and the bond valence sum for calcium in the top calcite monolayer was: BVS(Ca) = 2.17.

4. Results and discussion

The structural fitting parameters are listed in Table 4.7.

Tab. 4.7: Structural parameters obtained from fitting of dataset G. Uncertainties are twice the standard deviation calculated by the ROD software. Displacement parameters Δx, Δy, and Δz are in fractional coordinates, B_j are Debye–Waller factors in Å^2. θ and ϕ are the carbonate rotation angles in degrees (see Figure 3.6 for definition). (fix) values have not been adjusted in the fit.

	Δx (b_1)	Δy (b_2)	Δz (b_3)	B_j
	structured water			
H$_2$O(2)	-0.017 ± 0.006	-0.13 ± 0.03	0.021 ± 0.005	10 ± 2
H$_2$O(1)	-0.002 ± 0.005	-0.19 ± 0.01	0.013 ± 0.004	8.7 (fix)
	1st calcite layer			
Ca	-0.002 ± 0.002	-0.004 ± 0.039	-0.003 ± 0.002	2 (fix)
CO$_3$	0.000 ± 0.004	0.006 ± 0.012	0.001 ± 0.004	6 (fix)
	$\theta = -4.3 \pm 1.7$,	$\phi = -0.0 \pm 1.4$		
	2nd calcite layer			
Ca	-0.001 ± 0.002	-0.004 ± 0.009	0.000 ± 0.001	0.7 (fix)
CO$_3$	0.000 ± 0.003	0.002 ± 0.008	0.000 ± 0.003	1.0 (fix)
	$\theta = -0.3 \pm 0.4$,	$\phi = -0.2 \pm 1.3$		

The molecular surface structure according to the parameters listed in Table 4.7 is illustrated in Figure 4.15.

In data acquisition for datasets B – D between 486 and 590 data points were collected. In datasets E – G the counting time increased by a factor of two. To adjust the overall time needed for collection of a dataset, the number of data points has been decreased to 230 – 283. The ROD software weights the importance of data points for the structure refinement by their uncertainty. Therefore differences in counting time might effect the comparability of structural parameters. For datasets E and F the approach to count longer and reduce the number of data points has been very successful, the data have smaller uncertainties compared to datasets B – D. The possibility to refine the interface structure has not been influenced by the smaller number of data points. In dataset G the (13L) rod has not been measured. Additional loss of data points due to contributions of higher harmonics of the Bragg peaks to the signal make the dataset incomplete. Therefore the uncertainties ob-

4. Results and discussion

tained for the structural parameters on this dataset are very big. Fitting of Debye–Waller factors resulted in unreasonable values, too high in the calcite layers, and too low in the first water layer. Therefore Debye–Waller factors are set to values similar to those of the other datasets for the structure refinement.

4.1.3. Discussion of the surface structure

Frequent checks if adjustment of occupancy factors result in an improvement of the fit resulted in values not significantly different from one (1.00 ± 0.08). Therefore, these values have been set to one for all surface atoms and water molecules (oxygens) in the final fits. The large variation in surface roughness, β (0.01 – 0.23), corresponding to a root mean square roughness of σ_{rms} (0.6 Å – 3.8 Å) is difficult to explain. The real roughness of the cleaved calcite sample is expected to be low, as found for datasets B, E, F, and G (β: 0.01 – 0.06; σ_{rms}: 0.6 Å – 1.7 Å). The roughness is not expected to change during the measurements in equilibrium solutions and high pH non–equilibrium solutions, where dissolution kinetics are extremely slow. The elevated roughness values in datasets C and D might be related to systematic errors due to variations in the data integration procedure or to a contamination of the sample with suspended calcite powder during injection of solution C.

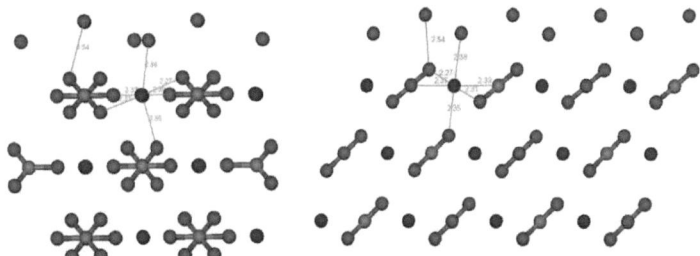

Fig. 4.15: Molecular structure of the calcite–water interface according to the result of the structure refinement of dataset G. The left hand side shows a projection along the $\mathbf{b_1}$ direction, the right hand side shows a projection along the $\mathbf{b_2}$ direction. Selected interatomic distances are indicated. (blue: Ca, gray: C, red: O)

4. Results and discussion

Some of the structural parameters do show significant variations throughout the different datasets. However, the similarities between the six structures far outweigh the differences. Let us consider the molecules at the interface one by one. Water molecules in the third and fourth water layer above the surface have very large Debye–Waller factors associated with them, B_j: 60 – 120 Å^2. They should not be regarded as structured. Modeling this part of the structure just with two water molecules is, as mentioned above, most probably an oversimplification as this is the region where outer–sphere complexes are expected. In order to model the data in the low L region of the (00L) rod correctly, it is, however, advantageous to have some diffuse electron density above the second ordered water layer. The water molecules of the second water layer ($H_2O(2)$) are located above the outward oriented oxygen atoms of the surface carbonate groups in the initial model. In the refined structures the distances between carbonate oxygen and $H_2O(2)$ range from 2.47 Å to 2.62 Å, with an average of 2.52 ± 0.06 Å [1]. Compared to the initial position directly above the carbonate oxygen, the water molecules are shifted along $\mathbf{b_2}$ by 0.5 ± 0.2 Å. This shift makes them line up midway between the carbonate oxygens in the projections along the $\mathbf{b_1}$ direction (cp. Figures 4.5 – 4.15). No general trend is observable in the shift along the $\mathbf{b_1}$ direction. The average distance of this molecule to the surface, defined by the z–position of the surface Ca atom, is 3.24±0.06 Å. This is in excellent agreement with recent molecular dynamics calculations (3.2 Å [56]). According to these calculations this water molecule is bound to the surface carbonate oxygen via a strong hydrogen bond. Previous surface diffraction studies on calcite find this water molecule at similar heights above the surface: 3.1 ± 0.1 Å [51] and 3.45 ± 0.16 Å [49]. Debye–Waller factors for this molecule vary between 10 and 20 Å^2. Increased Debye–Waller factors in datasets C and D might be related to the presence of outer–sphere calcium complexes that disturb the order in that water layer. However this effect has not been observed on dataset G, but as mentioned above, the Debye–Waller factors obtained for dataset G are probably not reliable.

The water molecules of the first water layer ($H_2O(1)$) are in the initial model located above the surface calcium at a position where in the bulk structure the oxygen of the next carbonate ion would be located. This initial constellation forms an ideal octahedron around the surface calcium atom with a Ca – O distance of 2.36 Å. In the refined structures the water molecule is not significantly translated in the $\mathbf{b_1}$ direction. In all the structures

---

[1]Uncertainties of average parameters given in this chapter are standard deviations of the values of the different datasets.

4. Results and discussion

$H_2O(1)$ is significantly shifted along b_2. The average shift is 1.1 ± 0.1 Å. This distorts the coordination polyhedron of the surface Ca considerably as can be seen in Figures 4.5 – 4.15 (the green lines indicate the calcium coordination polyhedron). The height above the surface changes only slightly compared to the initial position. The average translation along b_3 is 0.18 ± 0.12 Å. This results in an average height above the surface of 2.35 ± 0.05 Å. This value is again in good agreement with MD simulations (2.2 Å [56]). Literature values from surface diffraction studies vary in this respect. Geissbühler et al. (2004) find this molecule at 2.3 ± 0.1 Å [49] above the surface, which is in excellent agreement with this study. Fenter et al. (2000) report a value of 2.50 ± 0.12 Å [48], which is 0.15 Å larger than this study, but also agrees well. In the work of Magdans (2005), $H_2O(1)$ is found 1.9 ± 0.1 Å above the surface [51]. In this work the specular CTR has not been included in the structure refinement, this might explain the large deviation in the height. Debye–Waller factors of $H_2O(1)$ range from 8.7 – 11.5 Å^2.

No general trend can be observed for the shift of the position of the surface calcium atom, but shifts are very small throughout all datasets. In average the surface calcium is approximately at its bulk position.

The distance between surface Ca and $H_2O(1)$, which is the important parameter for bond valence calculations, varies between 2.33 Å and 2.49 Å. The average value is 2.41 ± 0.05 Å. This value is slightly larger than the distance in the ideal Ca – O octahedron (2.36 Å). Magdans finds the same distance: 2.4 ± 0.1 Å [51]. Geissbühler reports a very large value for this distance due to large lateral shifts: 2.97 ± 0.12 Å [49]. This large distance would correspond to an extremely weak interaction between surface calcium and the water molecule $H_2O(1)$. Our shorter distance is in much better agreement with MD simulations that report direct bonding between the water oxygen and the surface calcium [56].

A slightly enlarged distance between surface calcium and $H_2O(1)$ makes the bond valence sum of surface Ca smaller than two. This is compensated by motion of the surface carbonate ion. Actually except for dataset B, it is even slightly overcompensated. The carbonate ion shows an average shift in the b_2 direction about 0.03 ± 0.01 Å. In dataset B it is shifted in the opposite direction. The changes of the parameters during the fitting procedure suggests that this difference is due to additional degrees of freedom in dataset B resulting from the lack of data between $L = 0$ and $L = -1.7$ on the (13L) rod. The surface carbonate ion is tilted towards the surface (angle θ) by $4° \pm 1°$. The average ϕ rotation is not significantly different from zero $0° \pm 1°$. Debye–Waller factors of the surface carbonate

4. Results and discussion

are for all datasets about three times as high as for the surface calcium. Average values are 5.8 ± 0.4 Å^2 and 1.7 ± 0.5 Å^2, respectively.

The displacement parameters and tilt angles in the second calcite layer are very small.

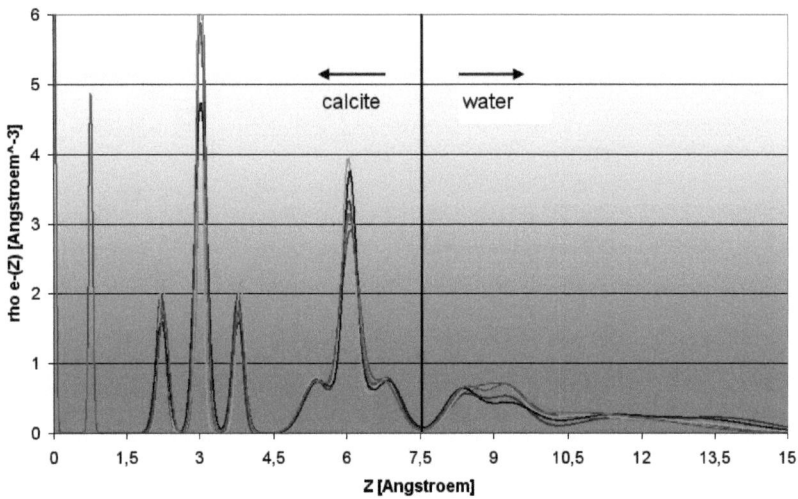

Fig. 4.16: Electron density distribution, $\rho_{e^-}(z)$ (Å^{-3}), in the surface unit cell according to all six structures projected onto the $\mathbf{b_3}$ vector ($z \cdot \frac{\mathbf{b_3}}{|\mathbf{b_3}|}$ (Å)). B: dark green, C: orange, D: blue, E: purple, F: green, G: red

No general trends can be observed, and the average values are not significantly different from zero. Debye–Waller factors lie as expected between the values of the first layer and the bulk values; the difference between carbonate and calcium is not as pronounced as in the first layer. Average values are, 0.7 ± 0.5 and 1.0 ± 0.4 for calcium and carbonate, respectively.

A good way to compare the different structures directly is by plotting their corresponding electron density distributions projected onto the $\mathbf{b_3}$ vector. Such plots are shown in Figure 4.16. The two peaks below 1.5 Å belong to the third calcite layer that has bulk structure. The peaks between 1.5 Å and 4.5 Å belong to the second calcite layer. The peaks between 4.5 Å and 7.5 Å belong to the first calcite layer. The high peak in the middle of each

4. Results and discussion

calcite layer is related to electron density from calcium, carbon, and the in plane carbonate oxygen, the two smaller ones on either side are electron density from the out of plane carbonate oxygen atoms. The calcite terminates about 7.5 Å. Between 7.5 Å and about 10 Å contributions from the first and second water layer can be seen. The broad electron density distribution above 10 Å is related to the not well ordered third and fourth water layer.

The most obvious differences between the datasets concerning the calcite layers are related to peak heights and widths. As the site occupancy is one for all surface atoms these are uniquely related to different Debye–Waller factors. The positional variations are hardly visible.

The slight variations in the heights of the water layers above the surface are visible. An interesting point is the transition between the second water layer and the rather diffuse electron density related to the third and fourth water layer. According to MD simulations a drop in water density above the second water layer is expected. In these simulations water density drops beyond the second water layer at 3.2 Å ($\sim$ 9.2 Å in Figure 4.16) and peaks up again in a third water layer at about 5 Å ($\sim$ 11 Å in Figure 4.16) above the surface [56].

In this distance from the surface outer–sphere complexes or water molecules from the hydration sphere of outer–sphere complexes might also be expected. The diffuse electron density that is generated by the third and fourth water layer might as well be associated with highly disordered outer–sphere complexes. The region on the CTRs that is mainly influenced by scattering from the rather diffuse electron density above the second water layer is, as already mentioned, the low L region (L < 1.5) on the (00L) rod. This is also the region where the differences between the datasets are most likely to be "real". A drawback for structure refinement is that a large error related to background subtraction is associated with that region, too. The many degrees of freedom in a fit model and the high Debye–Waller factors associated with atoms that high above the surface (B_j: 60 – 120 Å^2) additionally complicate the data analysis in that region. Still it might be possible that the differences in the electron density distribution at about 10.5 Å in Figure 4.16 reflect real differences between the datasets regarding outer–sphere complexation. Especially the fact that electron density distributions related to datasets measured in calcium containing solutions show flat transitions, while in the electron density distribution related to dataset E, the only one measured in calcium free solution, shows a drop in electron density between

4. Results and discussion

Z = 10 Å and Z = 11 Å is rather intriguing.

To summarize the results of the structure analysis:

- The part of the structure of the calcite(104)–water interface, surface diffraction measurements are mainly sensitive to, does not change significantly, even upon the investigated extreme changes in the composition of the contact solution.

- No indication for calcium or carbonate inner–sphere complexes on the flat calcite (104)–face has been observed.

- Two well ordered layers of water have been identified at 2.35 ± 0.05 Å and 3.24 ± 0.06 Å above the calcite surface. Water molecules of the first layer are located above the surface calcium ions and those of the second layer are located above the surface carbonate ions.

- In contact to solution mainly surface carbonate ions relax slightly from their bulk position and tilt towards the surface by about 4°.

- A gradual increase in Debye–Waller factors from the bulk crystal to the solution has been observed.

- Elevated Debye–Waller factors of water molecules in the second water layer in datasets C and D as well as the drop in diffuse electron density at about 4.2 Å above the surface in dataset E might be indicative for calcium outer–sphere complexes.

4.1.4. A new calcite surface complexation model

Based on the results of the surface diffraction experiment the question arises if it is possible to model the measured zetapotentials by a model that includes only outer-sphere adsorption of calcium, carbonate, bicarbonate, sodium, and chloride. Directly at the surface (in the 0–plane) only protonation and deprotonation reactions should be allowed.
To test this, the Basic–Stern SCM described in section 3.7 has been set up, and the parameters have been adjusted to fit the measured zetapotentials.
Best fit parameters are :

4. Results and discussion

- $K_1 = -0.17$ ($>$CaOH$^{(x-1)}$ + H$^+$ $\rightleftharpoons$ $>$CaOH$_2^x$)

- $K_2 = 0.17$ ($>$CO$_3$H$^{(1-x)}$ $\rightleftharpoons$ $>$CO$_3^{-x}$ + H$^+$)

- $C_1 = 0.45$ F/m^2

- $x_1 = 0.33$ nm (slip plane distance at IS = 0.1 M)

- $x_2 = 0.14$ nm (slip plane distance at IS = 0.05 M)

- $x_3 = 2.94$ nm (slip plane distance at IS = 0.01 M)

- $x_4 = 13.2$ nm (slip plane distance at IS = 0.001 M)

- IB(Na$^+$) = 0.56 ($>$(-) $\rightleftharpoons$ $>$(-) $\cdots$ Na$^+$)

- IB(Ca^{2+}) = 1.68 ($>$(-) $\rightleftharpoons$ $>$(-) $\cdots$ Ca^{2+})

- IB(Cl$^-$) = -0.05 ($>$(+) $\rightleftharpoons$ $>$(+) $\cdots$ Cl$^-$)

- IB(HCO$_3^-$) = 0.04 ($>$(+) $\rightleftharpoons$ $>$(+) $\cdots$ HCO$_3^-$)

- IB(CO$_3^{2-}$) = 3.26 ($>$(+) $\rightleftharpoons$ $>$(+) $\cdots$ CO$_3^{2-}$)

According to bond valence calculations the effective charge of the surface groups, x, should be about 0.25. The fact that $K_1 = -K_2$ results in an equal abundance of $>$CaOH$^{(x-1)}$ and $>$CO$_3^{-x}$ on the one hand and $>$CO$_3$H$^{(1-x)}$ and $>$CaOH$_2^x$ on the other hand. As $(x-1) + (-x) = -1$ and $(1-x) + x = 1$, the value of x does not affect the model. Villegas–Jimenez et al. [63] have observed an equal abundance of corresponding species at the $>$CaOH$^{(x-1)}$ and $>$CO$_3$H^{1-x} sites, too, and therefore proposed to treat them as one generic $>$CaCO$_3$ site. In princlple the observation of the equal abundance of corresponding surface species at the two sites is in perfect agreement with the best fit values of K$_1$ and K$_2$. The problem with the generic $>$CaCO$_3$ surface site is, however, that it leaves no charged sites at the surface above which outer–sphere complexes can be placed.

Values of K$_1$ and K$_2$ close to zero result in a weak dependence of the surface charge on pH in the pH range investigated (5.5 – 11). They cause the zero plane to be negatively charged in the whole experimental pH region. This offers a simple description of the charging behavior observed in non–equilibrium solutions. In Figure 4.17 the relative surface proton charge calculated by the SCM for solutions in equilibrium with air and for the titration

4. Results and discussion

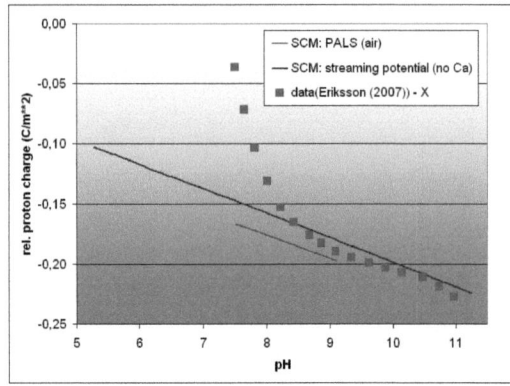

Fig. 4.17: Comparison between literature data on calcite surface proton charge [115] and corresponding SCM calculations.

in absence of calcium is compared to corresponding data taken from literature [115]. A value X = 0.17 C/m^2 is subtracted from the literature data points to better match the model values. Experimentally only the relative change in surface charge with pH can be determined. Therefore the absolute position of the surface proton charge curve is arbitrary and the approach to adjust the values by adding a constant is legitimate.

Because of the buffering effect of carbonate and bicarbonate, solutions should be carbonate free to obtain reasonable results from experimental investigation of surface proton charge. Therefore proton charge titrations on calcite are error-prone from the start. The problems related to investigations of the calcite surface in non–equilibrium solutions discussed in section 4.1.1 apply as well for proton charge titrations and shed additional doubt on these experimental results.

Nevertheless, it is good to see how well the model fits with the experimental data at pH > 8. As to how grave the discrepancy between the SCM calculations and experimental data at pH < 8 actually is, depends a lot on the reliability of the experimental data, which especially in the lower pH region might be questionable.

Within the MUSIC model [116] bond distances have an enormous effect on the protonation constants of surface oxygen groups. According to the surface structures obtained from surface diffraction, the distance between the outward oriented carbonate oxygen and the surface calcium atom ranges from 2.26 Å to 2.31 Å. Thus the contribution of the surface

4. Results and discussion

calcium atom to the bond valence sum of this carbonate oxygen atom ranges from 0.40 v.u. to 0.45 v.u. (v.u. = valence unit). Additional contributions to its bond valence sum come from the carbon atom: 1.33 v.u. and from a hydrogen bonded water: 0.25 v.u. [58]. This results according to the MUSIC model in a value for K_2 between 0.66 and -0.33 which bridges the best fit value, $K_2 = 0.17$. Using the same approach on the water molecule above the surface calcium atom, where calcium–oxygen bond distances range from 2.33 Å to 2.49 Å, we find the contribution from the surface calcium to the bond valence sum of the water oxygen ranges from 0.37 v.u. to 0.24 v.u. The oxygen atom of the deprotonated water molecule is bound to one donating hydrogen (0.75 v.u.) and accepts one or two hydrogen bonds (0.25 v.u. or 0.50 v.u.). Correspondingly the MUSIC model predicts values for K_1 ranging from 7.5 to 15. This is orders of magnitude higher than the best fit value, $K_1 = -0.17$.

As already mentioned in section 2.2.2 one of the major drawbacks of previous constant capacitance SCMs for calcite [60, 88, 62] are unrealisticly high capacitance values $\sim 10 - 100$ F/m^2, in addition to describing all mineral ion interactions as inner–sphere. To estimate reasonable values for the surface capacitance, the thickness of the Stern layer and the permittivity of the surface water molecules is needed. Relating the Basic–Stern model to the results of the surface structure refinement the 0–plane, where protonation and deprotonation reactions occur is located between the outward oriented surface carbonate oxygens and the first water layer. This location is about 1.5 Å above the surface calcium atoms. The b–plane, where outer–sphere complexation is assumed to take place, is expected beyond the second water layer at about 4 Å – 6 Å above the surface calcium atoms. This results in a approximate Stern layer thickness of about 3.5 Å. The permittivity of the well structured surface water molecules should as a first approximation be between that of liquid water (78.5) and that of ice (6). Correspondingly, reasonable values for the inner Helmholtz capacitance should be in the range of 0.1 F/m^2 to 2.8 F/m^2. The capacitance $C_1 = 0.45$ F/m^2 is in line with this estimate. The permittivity of the surface water layers would accordingly be about 18.

Slip plane distances are according to the Gouy–Chapman relation expected to be proportional to the Debye length, $1/\kappa$ (for a definition of κ see section 3.5.1). Their relation to ionic strength, I, is then proportional to $1/\sqrt{I}$. Slip plane distances are not measured from the surface but from the b–plane (cp. Figure 3.9). In Figure 4.18 the slip plane distances fitted for the calcite SCM are compared to those of SCMs for various other surfaces

4. Results and discussion

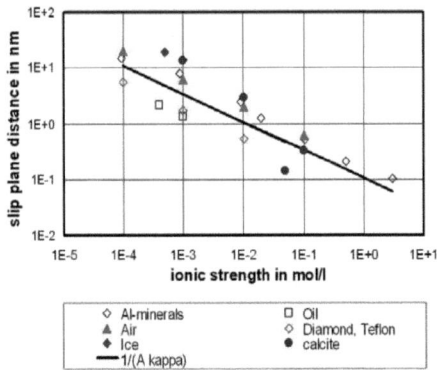

Fig. 4.18: Slip plane distances as a function of ionic strength. Calcite compared to various other surfaces (after [117]).

[117]. Regardless of the material the values for the slip plane distance as a function of ionic strength are all similar. The black line shows the relation: $x_i = A \cdot 1/\kappa$ with the proportionality factor $A = 0.0013$.

In the Basic–Stern model the calcite zetapotential is mainly determined by outer–sphere sorbed ions. The strength of outer–sphere adsorption of particular ions at the calcite surface is given by ion binding constants, IB. They describe the binding of an ion to a generic surface site of opposite charge. Monovalent anions adsorb only weakly (IB(Cl^-) = -0.05, IB(HCO_3^-) = 0.04). The monovalent cation sodium already shows a quite strong adsorption (IB(Na^+) = 0.56). The main contributions to the surface potential, however, come from strongly outer–sphere sorbing calcium (IB(Ca^{2+}) = 1.68) and carbonate (IB(CO_3^{2-}) = 3.26) ions.

Morel and Hering (1993) [118] present a theoretical expression to calculate stability constants of outer–sphere complexes. However, according to their expression the stability constants depend on many parameters that are not well constrained in the calcite–water interface system. For example, it is hard to define an ionic strength in the Stern layer. The distances between outer–sphere sorbing ions and the surface sites are not known and they might be different for all kind of outer–sphere complexes. The relative permittivity of the water in the Stern layer can be approximated from fit results for the Helmholtz

4. Results and discussion

capacitance, but there is a high uncertainty related to this parameter and it has a large influence on calculated stability constants for outer–sphere complexes. Additional uncertainty would originate in the fractional charges of the surface sites that would have to be taken into account.

Therefore trying to use the Morel and Hering expression to calculate the stability constants for outer–sphere complexes would result in more unknown parameters than simply using the ion binding constants as adjustable parameters. The theoretical expression can however be used to estimate reasonable limits for the ion binding constants; these are: $-0.3 \leq \text{IB} \leq 5$. The ion binding constants from fits lie within this range.

How the Basic–Stern SCM reproduces the experimental zetapotentials is shown in Figures

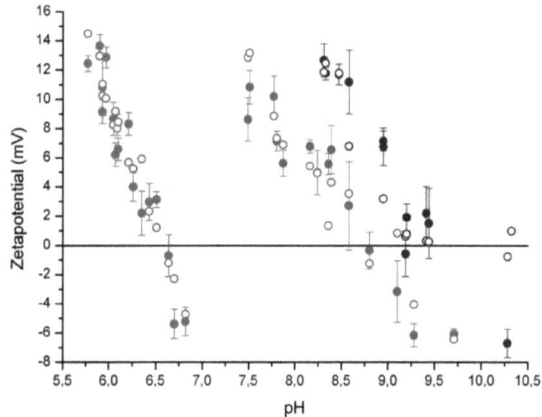

Fig. 4.19: Comparison of electrophoretic zetapotential data (full circles) and SCM calculations (open circles). Data is measured in equilibrium solutions at an ionic strength of 0.1 M NaCl at various solution compositions and CO_2 partial pressures. Red: equilibrium with CO_2, green: equilibrium with air (360 ppm CO_2), and blue: equilibrium with N_2 (6 ppm CO_2).

4.19 and 4.20. Only those streaming potential measurements that did not show hysteresis effects are considered for modeling (cp. section 4.1.1), as there is not yet a good explanation for this effect. A good explanation is also lacking for the increase in zetapotential at high pH and high calcium concentrations. There is not yet any reaction included in the

4. Results and discussion

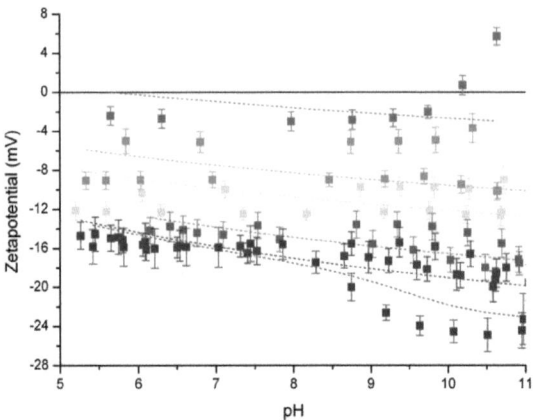

Fig. 4.20: Comparison of zetapotentials obtained by streaming potential measurements (squares) and SCM calculations (dashed lines). Background electrolyte concentration is 0.01 M NaCl. Purple: 5 mM Na_2CO_3, blue: no additional ions in solution, green: 0.1 mM $CaCl_2$, yellow: 0.5 mM $CaCl_2$, orange: 1 mM $CaCl_2$, and red: 5 mM $CaCl_2$.

model that could account for this behavior. Therefore the red data points in Figure 4.20 show the largest mismatch between model and data. Some data points in the PALS data at low CO_2 partial pressure (blue circles) are also not yet well described by the model, but generally the agreement between model and data is good. The main effects that have been observed in zetapotential measurements; the zetapotential in equilibrium solutions, the shift of the IEP with changing CO_2 partial pressure, the weak dependency of zetapotential on pH changes in non–equilibrium solutions, and the predominance of carbonate and calcium adsorption in the determination zetapotentials,are all well described by this Basic–Stern SCM that, according to surface diffraction observations, considers only outer–sphere complexes of ions other than protons and hydroxide.

Consistency with the results of the surface diffraction study has been the main focus in the development this Basic–Stern SCM. The degrees of freedom of the Basic–Stern SCM allow to describe the zetapotentials with reasonable values for the Helmholtz capacitance, the ion–binding constants and the slip plane distances.

4. Results and discussion

On the other hand, due to doubts about the reliability of the results of proton charge titrations on calcite, no data specific for inner–sphere protonation reactions has been used for the calibration of the model. The reaction constants K_1 and K_2 have not been constrained; they are merely fitting parameters. The value obtained for $K_2 = 0.17$ is nevertheless in excellent agreement with a MUSIC model prediction. The value for $K_1 = -0.17$ cannot be explained by the MUSIC model. However, it offers a simple way to describe the low zetapotentials observed over the whole pH range by streaming potential measurements in pure NaCl solutions.

The value of K_1 and the unreliability of experimental data on calcite obtained in non–equilibrium solutions in general are the main sources of uncertainties linked with this Basic–Stern SCM for calcite. Advantages are the good description of the zetapotential data over large ranges of experimental conditions and the consitency with structural observations and theoretical considerations discussed above.

Differences to other SCMs for calcite are mainly related to the different priorities set by their authors and judgment of the quality of the models is hardly possible. The Constant Capacitance Model by Pokrovsky an Schott [62] is a very simple model. It describes all interactions as inner–sphere even though the authors report that calcium forms most probably outer–sphere complexes. The unreasonably high Helmholtz capacitance is interpreted as indication for a collapsed electric double layer regardless of structural studies about the calcite–water interface. A newer version of this SCM has been publish recently summarizing calcium and carbonate surface species to one generic >$CaCO_3$ site [63]. The Constant Capacitance SCM is simple and nevertheless capable of describing many kinds of experimental data.

There is also very sophisticated three plane CD-MUSIC SCM for calcite [58]. Reaction constants of the surface sites are derived from the MUSIC equation [116] or defined in analogy to solution species' reactions, as in the first calcite SCM by van Cappellen et al. (1993) [60]. Inner– and outer–sphere complexes are considered and the Helmholtz capacitance is high, comparable to the Constant Capacitance SCM. The high complexity of the model is necessary to constrain the model parameters without loosing the capability to reproduce experimental data. One particular weakness of this model is that the bond distance between the surface calcium atom and the water molecule above it is taken from x–ray reflectivity results [48]: 2.50 ±0.12 Å. As x–ray reflectivity is only sensitive for the height of the oxygen above the surface and no lateral displacement is considered, the 2.50

4. Results and discussion

±0.12 Å is only a bond distance in the special case, where the lateral displacement is zero. The bond distance value, however, has a tremendous impact on potentials calculated with the model [58].

4.1.5. Inner–sphere adsorption at step sites?

No indication for inner–sphere complexes of calcium, carbonate, sodium, or chlorine has been found in the surface diffraction investigation of the calcite (104)–face. Surface diffraction is, however, only sensitive to processes causing structural changes at flat terrace planes on the crystal face. In situ atomic force microscopy (AFM) precipitation and dissolution experiment demonstrate that the most reactive sites at the calcite (104)–face are step and kink sites (cp. sections 2.2 and 2.3). These are the sites were crystal dissolution mainly proceeds. Crystal growth at low supersaturation takes place only at step and kink sites and not by adsorption of single ions on the ideal surface. In principle crystal growth at low supersaturation is not different from repeated sequential inner–sphere adsorption of the crystal constituents at steps on the surface. Therefore, assuming that inner–sphere adsorption of the crystal constituents might take place at step sites seems likely. In Figure 4.28 (section 4.2.3) an AFM image of a freshly cleaved calcite surface is shown that exhibits a density of step sites of about 0.5 %. It is difficult to predict how much influence these sites have on the zetapotential, but it might be possible that some of the effects which are not yet well described by the Basic–Stern SCM are caused by reactions taking place at step sites on the calcite (104)–face. Inclusion of step sites into the model might be an important next step to improve the Basic–Stern SCM as a tool to calculate the calcite surface charge. In the next section we will see that this might as well be crucial to enable model predictions about neptunyl adsorption at calcite. Mathematically this is quite simple and attempts in this direction have already been made during development of the SCM presented here. The complicated part is to secure reliable experimental data, within which the reaction constants of protonation and inner–sphere adsorption reactions at the step site calcium and carbonate ions can be calibrated.

4.2. Neptunyl(V) adsorption to calcite

4.2.1. Adsorption isotherms

Four adsorption isotherms have been derived from the adsorption experiments at four different CO_2 partial pressures in a concentration range between 0.4 and 40 μM NpO_2^+. The isotherms follow the empirical Freundlich relation between surface loading, q (mol/g), and concentration, c (mol/L), $q = K_F \cdot c^n$. K_F and n are empirical constants derived from linear regression of $\log_{10}(q)$ versus $\log_{10}(c)$. Results of the adsorption experiments are summa-

Tab. 4.8: Experimental conditions (CO_2 partial pressure and resulting pH) and Freundlich parameters resulting from the adsorption experiments. I = 0.1 M NaCl, calcite content = 20 g/L, 0.4 μM $\leq c_0(NpO_2^+) \leq$ 30 μM (40 μM at pH 8.3). Uncertainties in parenthesis are standard deviations of the linear regressions.[119]

$\log_{10}(p(CO_2))$	pH	K_F	n
0 (CO_2)	6	8.0($\pm$ 0.4)$\cdot 10^{-5}$	0.578($\pm$ 0.005)
-1.7 (mix)	7.2	5.2($\pm$ 1.2)$\cdot 10^{-5}$	0.531($\pm$ 0.020)
-3.44 (air)	8.3	2.4($\pm$ 0.4)$\cdot 10^{-5}$	0.464($\pm$ 0.014)
-5.2 (N_2)	9.4	6.2($\pm$ 1.2)$\cdot 10^{-5}$	0.544($\pm$ 0.017)

rized in Table 4.8 and shown in Figure 4.21. Maximum adsorption is at pH 8.3. At low concentrations, the pH dependence is stronger than at higher concentrations. Above 10 μM NpO_2^+ in equilibrium solution there is no significant pH dependence. Freundlich-isothermic adsorption behavior means that it is not possible to describe adsorption by one simple KD value (KD = q/c). In this system, KD is a function of pH and concentration. The decrease of adsorption with increasing concentration cannot be explained with a saturation of the available surface sites. In our experiments with 20 g/L calcite in suspension, with 1.28 m^2/g specific surface area, and a site density of 8.22 μmol/m^2 (= 4.95 nm^{-2}) there are 210 μmol/L surface sites available. Which is an excess of surface sites. The highest loading of 0.213 μmol/g corresponds to 4.23 μmol/L in our experiments, which translates to a surface site occupancy of only 2%. Zavarin et al. (2005) [28] reported on NpO_2^+ adsorption to calcite. Their experimental results agree very well with those presented here. Their adsorption experiments with an NpO_2^+ concentration of 0.1 μM at lower calcite concentrations (1.25 g/L) result in 25% or less adsorption, with a similar pH dependence. We find

4. Results and discussion

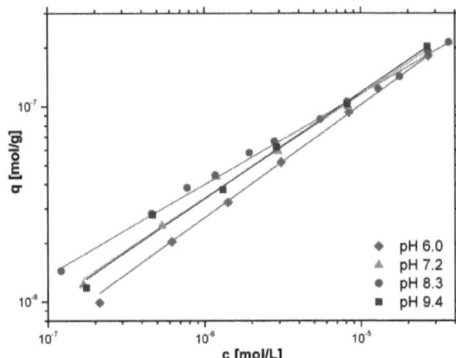

Fig. 4.21: Isotherms resulting from the adsorption experiments. Shown is surface loading, q (mol/g), versus solution concentration, c (mol/L). Measured values (symbols) and Freundlich-isotherms (lines) are displayed for the four pH values investigated. All experiments are carried out with c(NaCl) = 0.1 M, and a calcite content of 20 g/L.

70% or less due to the higher calcite content in the experiments. Calculated as percentage of the total NpO_2^+ in the system, the adsorbed fraction of NpO_2^+ increases with decreasing total NpO_2^+ concentration. This trend is also described by Keeney-Kennicutt and Morse (1984) [29]. In their study this trend carries on down to NpO_2^+ concentrations of about 10^{-8} M. Below this concentration the adsorbed fraction is constant at about 95%. Differences between adsorption isotherms in their study and ours are mainly caused by the different solid to solution ratios and the different reaction times.

4.2.2. Adsorption kinetics, desorption, and irreversibility

Kinetic adsorption experiments show that after 24 to 48 h adsorption slows down remarkably. Therefore reaction time for the determination of adsorption isotherms has been set to 72 h in this study. It is worth mentioning that even after 4 weeks reaction time adsorption does not reach an equilibrium state; between 2 days and 4 weeks a very slow but significant increase in surface loading is observed. The course of adsorption from initially 2 μM NpO_2^+ solutions is shown in Figure 4.22. Desorption experiments performed with

4. Results and discussion

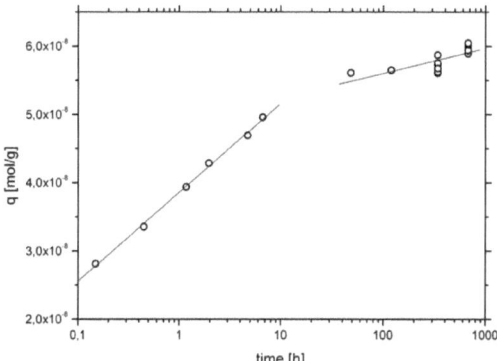

Fig. 4.22: Experiments on adsorption kinetics. Shown is surface load q (mol/g) versus time (h). Initial NpO_2^+ concentration is 2 µM, c(NaCl) = 0.1 M, and calcite content is 20 g/L.[119]

the calcite after 4 weeks of adsorption result in a KD value of 0.090±0.004 L/g after 5 h to 8 days. The KD value of the adsorption after 72 h at the corresponding pH and final concentration is 0.061±0.002 L/g. These different KD values indicate partial irreversible sorption. One possible explanation might be that NpO_2^+ becomes structurally incorporated at the surface. Similar observations have been reported in many studies of trace element adsorption at the calcite water interface (cp. section 2.4). If the hypothesis that NpO_2^+ becomes incorporated into calcite at calcite equilibrium conditions could be proven a thermodynamic driving force for the structural incorporation of NpO_2^+ into calcite must exist. This will be topic of section 4.3.

EXAFS measurements have been performed to search for evidence for recrystallization processes that lead to neptunyl incorporation into the calcite surface layers at calcite equilibrium conditions over extended reaction times. In experiments especially for the preparation of the EXAFS samples, a slight increase in the amount of adsorbed neptunyl is again observed between 48 hours and three months: 0.14 µmol/g and 0.21 µmol/g of adsorbed neptunyl in samples Np-Y1 and Np-Y2 (48 h) and 0.27 µmol/g adsorbed neptunyl in sample Np-O2 (three months). However, no significant difference in the EXAFS spectra of samples Np-Y1 and Np-O2 is visible (cp. Figure 4.23), except for the better

4. Results and discussion

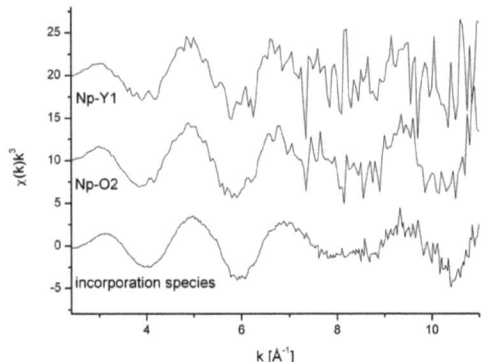

Fig. 4.23: k^3-weighted EXAFS data of samples Np-Y1, Np-O2, and for comparison an example for a spectrum of an incorporation species from a mixed flow reactor experiment [27]. Spectra are offset along the ordinate for clarity.

signal to noise ratio in sample Np-O2 due to the higher neptunyl content in this sample. This means there is no spectroscopic evidence for neptunyl incorporation into calcite in batch experiments at calcite equilibrium conditions.

The spectra of adsorption and incorporation species are quite similar (cp. Figure 4.23). If the incorporation species exists it would make only a few percent of the neptunyl content of the EXAFS sample Np-O2 (cp. Figure 4.22). We may still argue that incorporation takes place and the resulting incorporation species is not distinguishable from EXAFS data.

Due to the better quality of the spectrum measured on sample Np-O2 this spectrum has been used for a detailed analysis of the adsorption complex structure.

4.2.3. The adsorption complex structure

The k^3-weighted extracted EXAFS signal of sample Np-O2 together with the model spectrum is shown in Figure 4.24 a). Figure 4.24 b) displays the corresponding Fourier transform spectra in R-space. A structural model giving a good fit between measured and modeled spectrum and saturating the bond-valence of the neptunium atom has been achieved using five different electron backscattering paths. Structural parameters obtained from the

4. Results and discussion

EXAFS data analyses are listed in Table 4.9. However, it is not trivial to find a structural explanation for the measured model parameters. The 2.0($\pm$0.1) oxygen backscatterers at

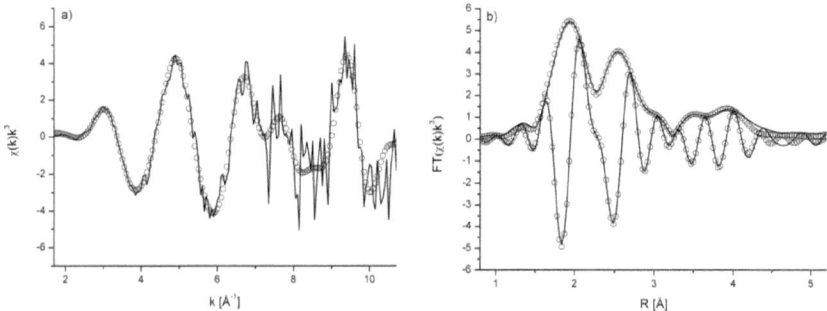

Fig. 4.24: a): k^3-weighted extracted EXAFS data (line) and model spectrum (circles) of sample Np-O2. b): Fourier transform amplitude and imaginary part of the k^3-weighted EXAFS data (line) and model spectrum (circles). Spectra are corrected for phase shift.

1.87($\pm$0.01) Å are surely related to the axial oxygen atoms (O-ax) of the linear neptunyl molecule. 6.1($\pm$0.5) oxygen backscatterers at 2.51($\pm$0.01) Å are associated with equatorial oxygen coordination shell (O-eq). This coordination environment is in good agreement with bond valence calculations. Using the bond valence parameters from Forbes et al. [105], R_0 = 2.035 and b = 0.422, the bond valence sum for the central neptunium atom BVS(Np) = 4.94.

The six equatorial oxygen neighbors at 2.51 Å and about three carbon backscatterers at 2.94 Å distance from the central neptunium atom can be explained by three bidentate bound carbonate groups. Bond distances and coordination numbers are in almost perfect agreement with those reported for the aqueous neptunyl triscarbonato complex: 6–7 Np–O at 2.53 ($\pm$0.03) Å and 2.7 ($\pm$0.6) Np–C at 2.98 ($\pm$0.03) Å [113]. The structure of this complex is shown in Figure 4.25.

However, from a chemical perspective formation of the neptunyl triscarbonato complex makes hardly any sense. Speciation calculations show that at the experimental conditions the free neptunyl ion and the neptunyl-monocarbonate complex dominate the speciation. The presence of two calcium backscatterers at 3.95($\pm$0.03) might suggest that neptunyl might form calcium–neptunyl–triscarbonato complexes analogous to the uranyl(VI) species

4. Results and discussion

Tab. 4.9: Coordination numbers, N, distances, R (Å), and Debye-Waller factors, σ^2 (Å^2), from EXAFS data analyses. Uncertainties, in brackets, are twice the standard deviation calculated by the Artemis software. The amplitude reduction factor has been held constant at $S_0^2 = 0.8$, relative shift in ionization energy ΔE_0 is 8 ($\pm$1) eV, and the goodness of fit parameter for all three k-weightings together, r = 0.007. Number of independent points, NI = 20, and 16 adjustable parameters have been used to fit the data.

Backscattering path	N	R (Å)	σ^2 (Å^2)
O (O-ax)	2.0($\pm$0.1)	1.87($\pm$0.01)	0.001($\pm$0.001)
O (O-eq)	6.1($\pm$0.5)	2.51($\pm$0.01)	0.009($\pm$0.003)
C (C-bi)	2.7($\pm$1.2)	2.94($\pm$0.02)	0.001($\pm$0.003)
O (O-surf)	2.0($\pm$0.8)	3.50($\pm$0.04)	0.001($\pm$0.006)
Ca	1.8($\pm$1.0)	3.95($\pm$0.03)	0.003($\pm$0.004)

reported in literature [120]. These uranyl complexes form in calcium containing solutions even at chemical conditions, where the triscarbonato complex is not stable in corresponding calcium free solutions. Such a complex could adsorb at step-edges on the surface, similar to the mechanism that has been suggested by Reeder et al. [22] to explain the site specific incorporation of uranyl into calcite.

There are several points based on the EXAFS data that argue against such a complex. First, this complex does not explain the two oxygen backscatterers at 3.50($\pm$0.04) Å that make a significant contribution to the EXAFS. Second, spectra of the supernatant of adsorption experiments at the same experimental conditions do not show any evidence for this kind of complex [119]. Third, actinyl triscarbonato complexes usually show a very characteristic feature in the k-space spectrum between 6 and 8 Å^{-1} [14] that is not present in the sample Np-O2 spectrum.

Another possibility to explain the model parameters is to assume a split equatorial coordination shell consisting of two monodentate, and two bidentate bound carbonate groups. Indications for a split equatorial oxygen shell has also been found for uranyl adsorbed at calcite [21]. A split equatorial oxygen shell also offers a good explanation for the elevated Debye–Waller factor, 0.009($\pm$0.003) Å^2 that has been measured for the O-eq backscattering path. The static disorder inherent to a split shell would necessarily effect a large Debye–Waller factor. Indeed the second oscillation in the R-space spectrum can be well fit with two oxygen backscattering paths. If one oxygen backscattering path containing

4. Results and discussion

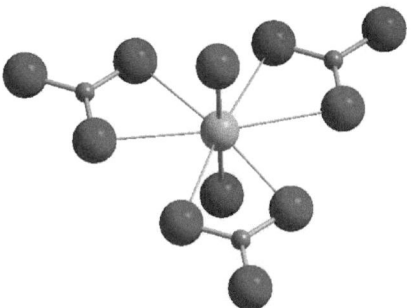

Fig. 4.25: Structure of the neptunyl triscarbonato complex with three bidentate bound carbonate ions. This structure offers a simple explanation for many of the results from the EXAFS analysis, but some facts contradict such a complex.

two atoms is assumed at a distance, R, and a second one containing four oxygen atoms is assumed at a distance R + 0.14 Å the two paths can be modeled with only the one adjustable radius, R, and one Debye-Waller factor. This results in an equally good fit and the Debye–Waller factor for these two paths decreases to the much more reasonable value of 0.002 Å^2[2]. The equatorial coordination would then have 2 oxygen atoms located at 2.39 Å distance and another four oxygens at 2.53 Å distance from the central neptunium atom. Bond valence sum for neptunium in this equatorial coordination is: BVS(Np) = 5.06. The split equatorial environment is therefore considered likely. Due to the limited k-range used in the fit, however, it is not possible to resolve such details from the EXAFS data and measure the radii of these backscattering paths individually, so this analyses cannot be considered a proof.

Let us, nevertheless, consider a split equatorial coordination with two monodentate bound surface carbonates and two bidentate bound carbonate ions and compare this to fit results. The two oxygen atoms located at 2.51 Å and two oxygen atoms at 3.50 Å originate surface carbonates bound to NpO$_2^+$ in a monodentate fashion. The two calcium atoms at 3.95 Å, also belong to the calcite surface. On the solution side four oxygen atoms are at a distance of 2.51 Å from the neptunium atom and two carbon backscatterers at 2.94 Å. These atoms

---

[2]Attention: Please note that the EXAFS Debye–Waller factor, σ^2, is not equivalent to the one used in surface diffraction, B. For isotropic Debye–Waller factors $B = 8\pi^2 \cdot \sigma^2$

comprise bidentate bound carbonate ions.

Basically three non–equivalent positions are available at the calcite (104)–face, where neptunyl can potentially adsorb in such a bidentate fashion. They are illustrated in Figure 4.26. Shown is a slab of (104)–terminated calcite. On the right, an obtuse step edge is

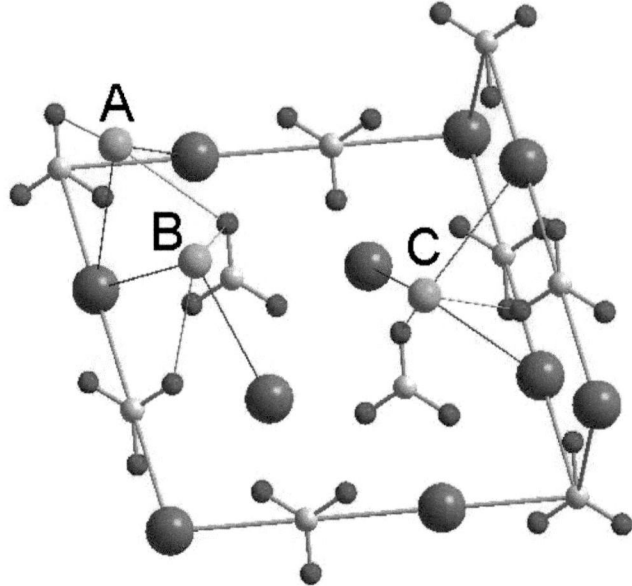

Fig. 4.26: Possible positions for bidentate adsorption of neptunyl at the calcite (104)–face. Position A and B at the flat calcite (104)–face Position C at a step edge site. Ca: cyan, O: red, C: gray, Np: green

indicated. Neptunium positions in Figure 4.26 are chosen in a way that the Np–Ca distances (blue lines) are 4 Å. The distances between Np and the closest two surface carbonate oxygens are equal (green lines). Positions A and B are located at the flat (104)–face and position C is at a step edge. Comparison of the resulting the distances between neptunium and the surface carbonate oxygen atoms associated with these hypothetical species reveals which position is in best agreement with the EXAFS results. Interatomic distances between Np located at position A and the surface oxygen atoms are 2×3.7 Å and 1×3.5

4. Results and discussion

Å; for position B they are 2 × 3.1 Å, 1 × 3.5 Å, and 1 × 3.9 Å. A tremendous distortion of the calcite surface structure would be necessary to achieve the bond distances measured with EXAFS. For position C at the step edge the distances are 2 × 2.36 Å and 2 × 3.44 Å. This is in excellent agreement with the EXAFS results. Only a slight relaxation of the surface around the neptunium is necessary to match the EXAFS bond distances. The only drawback of position C is that there are three instead of 1.8±1.0 calcium backscatterers. However, considering that error generally associated with EXAFS measurements is much greater for coordination numbers (∼ 20 %) than for bond distances (± 0.02 Å) a structure that matches the bond distances is preferred over one that matches the coordination numbers.

An obtuse step edge is shown in Figure 4.26 but the same structural arrangement is available at acute step edges. Accordingly the EXAFS results cannot be used to distinguish between the two.

A similar equatorial environment with two monodentate and two bidentate carbonate ions is reported for orthorhombic neptunyl monocarbonate phases: $MNpO_2CO_3$ (M = Li^+, Na^+) [40]. However, in these mineral phases the equatorial oxygen shell is strongly split and the Np–C bond distance is remarkably shorter: $R(Np-Oeq_{mono}) = 2.68$ Å, $R(Np-Oeq_{bi1}) = 2.43$ Å, $R(Np-Oeq_{bi2}) = 2.36$ Å, and $R(Np-Cbi) = 2.75$ Å, hence it is doubtful that the EXAFS data from sample Np-O2 represents a neptunyl carbonate precipitate rather than an adsorption complex.

This leads to the conclusion that the most abundant neptunyl adsorption species at the calcite (104)–face at the experimental conditions of this study most likely is adsorbed at step edges. There neptunyl forms a bidentate inner sphere sorption complex with two monodentate coordinating carbonate groups from the calcite surface. Two bidentate bound carbonate ions are located on the solution side of the neptunyl ion. A ball-and-stick representation of this sorption complex structure is shown in Figure 4.27.

The highest surface coverage of calcite with neptunyl found in adsorption experiments performed under the same experimental conditions as in the preparation of the EXAFS samples was 0.14 μmol/m^2. As already described above this corresponds to a surface site occupancy of less than 2%. Atomic force microscopy images of cleaved calcite surfaces support the idea that the site density of step edge sites on calcite grains could be in the order of magnitude of 1% (Figure 4.28).

It is not to be expected that the observed adsorption complex is the only adsorption

4. Results and discussion

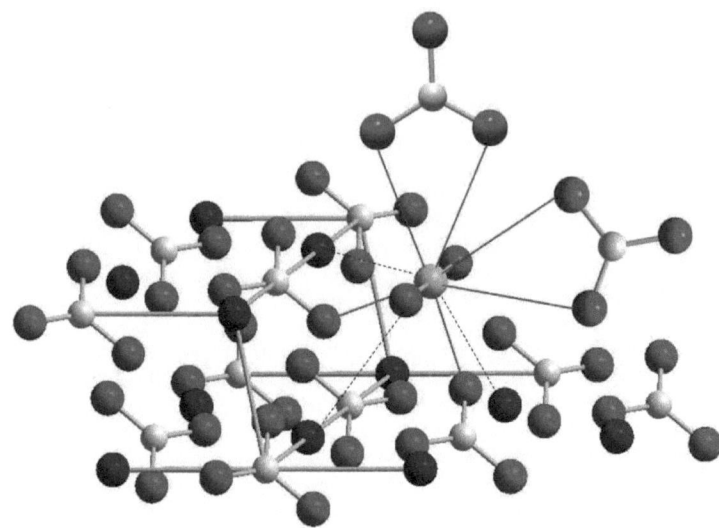

Fig. 4.27: Structure of the most likely neptunyl–calcite adsorption complex according to the results of the EXAFS data analyses: neptunyl sorbs at step edges on the calcite (104)–face as a bidentate inner–sphere biscarbonato complex. Indicated with green lines are the bonds to the six equatorial oxygen atoms. Dashed lines indicate the nearest calcium neighbors. (Np: green, Ca: blue, O: red, C: grey)

complex present at the calcite (104)–face across the whole pH and concentration range studied. It is likely that this is the most abundant and best ordered complex present under the experimental conditions the EXAFS sample has been prepared.

Support that several species can be present in adsorption samples while only one average structure is 'visible' for EXAFS is reported in literature. While EXAFS spectroscopy reveals only one adsorption complex structure at low surface coverages of uranyl adsorbed at calcite, time resolved laser fluorescence spectroscopy indicates that at least two species must be present [21].

4. Results and discussion

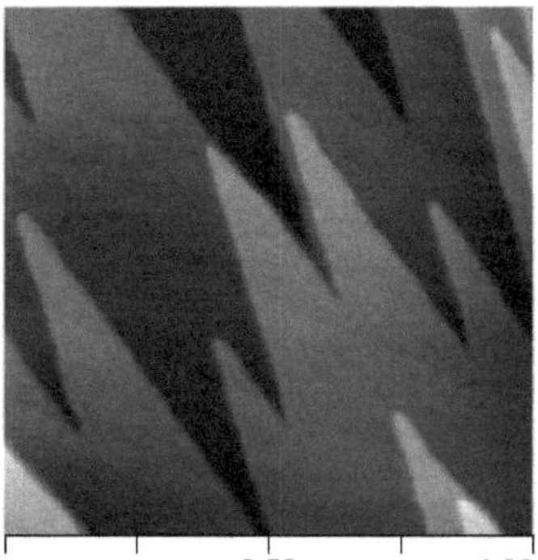

Fig. 4.28: AFM image of a freshly cleaved calcite (104)–face [73]. Visible are molecular steps (height 3.036 Å). Assuming 5 Å step width, about 0.5% of the 1 μm^2 large area of the 'flat' crystal face are covered with steps.

4.2.4. Inclusion of neptunyl adsorption at calcite into the SCM?

Adsorption SCMs can be used to quantify metal adsorption at mineral surfaces over a broad range of chemical conditions, e.g., pH, pCO_2, varying concentrations, etc. Because of its more universal applicability such a description would be highly preferred over description by KD values or empirical adsorption isotherms.

Zavarin et al. [28] published a model to describe neptunyl adsorption at calcite. Data and model calculation are introduced in section 2.4 and shown in Figure 2.8. Parameters to quantify the neptunyl adsorption at calcite surfaces are introduced based on the surface complexation model by Pokrovsky and Schott [62], using two monodentate neptunyl surface complexes: $>CaCO_3NpO_2$ and $>CaCO_3NpO_2CO_3^{2-}$. Their model describes the pH dependence of adsorption at one neptunyl concentration and solid to solution ratio used in their study.

However, attempts to compare the adsorption isotherms measured in this study with model predictions based on their reaction constants resulted in an overestimation of the adsorption by the model by more than five orders of magnitude. Their model parameters are also not able to describe adsorption data by Keeney–Kennicut and Morse (1984) [29]. Reaction constants by Zavarin et al. are obviously not intrinsic universal constants. Furthermore,

4. Results and discussion

the surface complexes they assume to model their data are not in agreement with the surface complex identified from EXAFS data in this study as the most abundant surface species.

As already mentioned in section 4.1.5 the inclusion of step sites into a SCM might be the key to successfully describe all effects observed during zetapotential measurements. According to EXAFS results presented here consideration of step sites in the SCM would also be necessary to enable inclusion of neptunyl adsorption at calcite into the SCM. To do this reliably is, however, a major challenge.

4.3. Neptunyl(V) coprecipitation with calcite

4.3.1. MFR experiments

In 11 MFR experiments, a 0.4-3.0 μm thick homogeneous calcite layer containing between 450 and 12000 ppm Np(V) has been synthesized onto calcite seed crystals at steady state conditions and precipitation rates between $4.4 \cdot 10^{-9} - 3.5 \cdot 10^{-8}$ mol/(m$^2 \cdot$ s). An empirical partition coefficient, D, (cp. equations 3.28, 3.29, and 3.30) is determined to be in the range from 0.5 to 10.3.

In experiments Np1 to Np4 the influence of pH on coprecipitation is investigated. The

Tab. 4.10: Steady state conditions, pH$_{out}$, and SI$_{out}$, steady state growth rate, R, and partition coefficient, D, resulting from the MFR experiments

Experiment	pH$_{out}$	SI$_{out}$	R ($\times 10^{-8}$ mol/(m$^2 \cdot$ s))	D
Cc1	10.13	0.04	1.2	–
Cc2	10.29	0.29	2.7	–
Cc3	10.32	0.22	2.8	–
Cc4	10.16	0.53	6.6	–
Np1	10.26	0.43	1.0	2.0
Np2	10.15	0.26	1.1	0.5
Np3	12.77	0.57	3.2	7.5
Np4	8.08	0.95	3.5	10.3
Np5	10.24	0.26	1.1	5.5
Np6	10.27	0.22	1.1	2.9
Np7	10.27	0.18	0.75	3.1
Np8	10.18	0.79	0.54	0.7
Np9	10.20	0.78	0.55	0.6
Np10	10.13	0.69	0.44	0.5
Np11(SC)	–	–	–	0.2
U1	7.80	0.59	6.8	0.02

pH has a major effect on the aqueous speciation of Np(V) [121]. At pH 10.2 the neptunyl monocarbonato complex dominates (90%) (Np1 + Np2). At pH 12.77 the neptunyl hydroxo

4. Results and discussion

and dihydroxo complexes are the major species (18 and 70%, respectively) (Np3). At pH 8.08 the neptunyl monocarbonato complex still dominates the aqueous speciation (76%), but the free NpO_2^+-ion makes about 24% of the Np content (Np4). However, there appears to be no clear correlation between D and pH (Table 4.10). If one substitutes the aqueous concentration ratio $c(Np(V))/c(Ca(II))$ by the activity ratio $\{NpO_2^+\}/\{Ca^{2+}\}$, D values in experiments Np1 to Np4 would then range from 5 (Np2) to 306 (Np3). This large range is not a consequence of the neptunyl content of the precipitates, rather it originates from the low NpO_2^+ activity at high pH where carbonato- and hydroxo-complexes dominate the speciation and free NpO_2^+ is hardly present. Therefore, we conclud that NpO_2^+ incorporation into calcite does not depend on the relative concentration of the NpO_2^+-aqueous species present in the experimental solution, but rather on the total Np(V) concentration.

The saturation state of the aqueous solution at which precipitation takes place is expressed as the output saturation index, SI_{out}. SI_{out} lies between 0.04 and 0.95 during the experiments. Under these conditions, supersaturated solutions remain metastable, and no homogeneous nucleation occurs within the time frame of the experiments [67]. In experiments Np1 to Np10 the coprecipitation rates are generally positively correlated with SI and vary between $4.4 \cdot 10^{-9} - 3.5 \cdot 10^{-8}$ mol/(m²· s). These rates are small compared to previously published values [122]. The presence of Np(V) in solution reduces the precipitation rate. This effect is shown in Figure 4.29, which includes four sets of MFR steady state growth rates versus steady state SI (SI_{out}). Pure calcite experiments Cc1 - Cc4 (orange data points) show the highest growth rates. In the neptunyl coprecipitation experiments (blue data points) growth is retarded. The biggest effect can be observed for experiments with the highest Np input concentration (Np8 - Np10, $c_{in}(Np) = 5$ μmol/L (blue triangles)). In experiments Np1 - Np4 ($c_{in}(Np) = 1$ μmol/L, blue diamonds) the growth rates scatter strongly around the regression line, probably due to the changes in pH in this experimental series. For experiments Np5 - Np8 ($c_{in}(Np) = 0.1$ μmol/L, blue squares) the growth retardation effect is as high as for experiments Np1 - Np4.

Such crystal growth inhibition effects are known for species in solution, which block active crystal growth surface sites (e.g., kink and step sites) [123, 35] (cp. section 2.5.2). This is in agreement with the adsorption complex structure suggested in section 4.2.3. Neptunyl adsorbs at calcite at step sites and retards the calcite crystal growth even at low face coverages.

In order to be able to compare our Np(V) coprecipitation experiments to previously pub-

4. Results and discussion

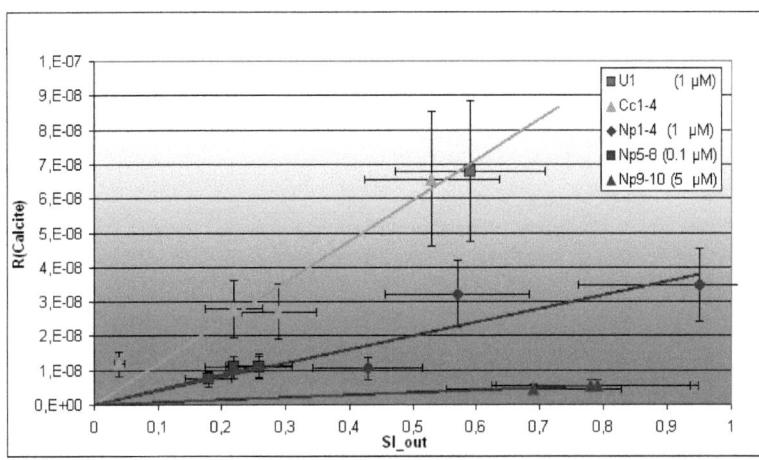

Fig. 4.29: Neptunyl coprecipitation with calcite retards the calcite crystal growth. Plotted is steady state growth rate, R, versus steady state SI, SI_{out}. Pure calcite experiments Cc1 -Cc4 (orange data points) show the highest growth rates. In the neptunyl coprecipitation experiments (blue data points) growth is retarded. The largest effect can be observed for experiments with the highest Np input concentration (Np8 - Np10, $c_{in}(Np) = 5$ μmol/L (blue triangles)). For experiments Np5 -Np8 ($c_{in}(Np) = 0.1$ μmol/L, blue squares), the growth retardation effect is as high as for experiments Np1 - Np4 ($c_in(Np) = 1$ μmol/L, blue diamonds). Uranyl coprecipitation (experiment U1, red square) does not cause retardation of crystal growth .

lished U(VI) coprecipitation experiments, one experiment has been performed with U(VI), U1, under similar conditions as, Np4 (SI_{in} 1.3 and pH_{in} 8.2). PhreeqC calculations show that there are no supersaturated uranyl phases under these conditions. From this experiment we derive a D of 0.02 (Table 4.10), corresponding to a 23 ppm U containing calcite. This value agrees well with previously published data [122, 24]. In comparison, the value for D in experiment Np4 is 2 orders of magnitude higher than for U1. This indicates Np(V) has a higher affinity for calcite than U(VI). The differences between the Np(V) and U(VI) coprecipitation behavior could be related to the high stability of U(VI) carbonato solution complexes. Under the conditions of experiment U1 the uranyl–calcium-triscarbonato complexes dominate the aquatic uranyl speciation (61% $Ca_2UO_2(CO_3)_3^0$ and

4. Results and discussion

39% $CaUO_2(CO_3)_3^{2-}$) whereas the monocarbonato complex and the free neptunyl ion dominate the Np(V) speciation (76 and 24%) in Np4. Assuming that this difference in solution speciation originates from a higher affinity of uranyl for bidentate complexation by carbonate ions offers a plausible explanation for the lower compatibility/affinity of the uranyl cation to the calcite structure. The crystal growth inhibition effect observed for Np(V) is not effective for U(VI) (see Figure 4.29).

4.3.2. Spectroscopic investigation of the incorporation sepcies

EXAFS characterization

As described in section 3.9.2 samples from experiments Np1 - Np4 and U1 have been used to characterize the structure around neptunyl and uranyl incorporated into calcite by means of EXAFS spectroscopy. The k^2-weighted EXAFS spectra (small diagrams) and their Fourier transform (FT) spectra in R-space are shown in Figure 4.30 together with the corresponding model (fit result) spectra. Spectra in R-space are not corrected for phase shift. Structural parameters obtained from the EXAFS analyses are summarized in Table 4.11. The distance from the central Np atom to the axial oxygens (Oax) is 1.86 (±0.01) Å. In the equatorial plane the neptunyl ion is coordinated by about four oxygen atoms (Oeq) at 2.40 (±0.01) Å distance. The coordination numbers for the more distant scattering paths (C/O2) are also about four for Np1 and Np2. The coordination numbers for these paths for samples Np3 and Np4 do not yield a stable fit so they are held constant at four during the fit procedure without any significant increase in the highly correlated Debye–Waller factor values. Measured Np-C distances range from 3.10 to 3.21 Å, R(Np-O2) from 3.35 to 3.45 Å.

In order to interpret these EXAFS analysis results, let us turn to the coordination structure of known Np(V) compounds. Pentavalent neptunium forms a linear NpO_2^+ complex in aqueous environments. Np-Oax distances range from 1.82 to 1.90 Å in solid neptunyl phases with equatorial oxygen coordination [124]. Atoms comprising the equatorial plane coordination show much stronger interatomic distance variability. Solid Np(V)-phases are known with four–, five–, and six–fold equatorial coordination in square, pentagonal, and hexagonal bipyramidal geometry [124]. In the case of oxygen as nearest neighbor in the equatorial plane, distances are near 2.39 Å for four–fold coordination, 2.39–2.52 Å for five–fold coordination, and 2.42–2.64 Å for six–fold coordination [124]. In the aqueous

4. Results and discussion

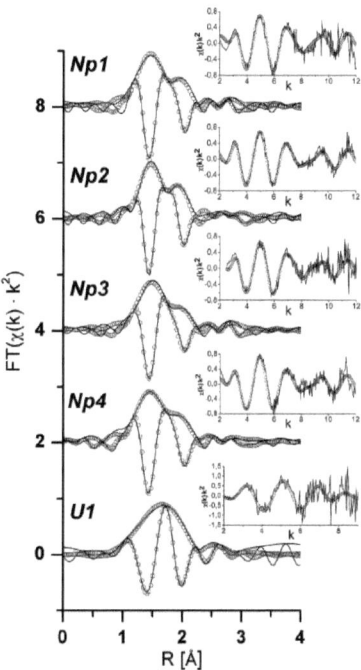

Fig. 4.30: Fourier transform amplitude and imaginary part of the k^2-weighted EXAFS data (lines) and model spectra (circles). Spectra are not corrected for phase shift and they are offset along the ordinate for clarity. The original EXAFS data in k-space together with the corresponding fits are shown in the small diagrams above each Fourier transformed spectrum.[27]

triscarbonato complex $(NpO_2(CO_3)_3^{5-})$ three carbonate ions coordinate Np(V) through bidentate bonds with a Np–Oeq distance of 2.53 Å [113]. Due to the bidentate charakter of the carbonate bonding, the Np–C distances are short (2.93–2.98 Å) [113] (cp. section 4.2.3, Figure 4.25). In solid neptunium monocarbonate phases $MNpO_2CO_3$ (M = Li^+, Na^+, K^+, NH_4^+, Rb^+, or Cs^+), the neptunyl ion is coordinated by six Oeq originating from either three bidentate or two mono and two bidentate carbonate groups, forming hexagonal or orthorhombic (NpO_2CO_3) layers depending on the size of the interlayer cation [40]. If

4. Results and discussion

Tab. 4.11: Results of the EXAFS analysis for coordination numbers, N, inter atomic distances, R, and Debye–Waller factors, σ^2. Uncertainties (in parentheses) are standard deviations given by the Feffit software. The uncertainties of the coordination numbers are increased about 10% to account for the uncertainty in S_0^2. (fix): Held constant during the fit. [27]

Np1	N	R (Å)	σ^2 (Å^2)
O_{ax}	2.1(±0.4)	1.86(±0.01)	0.0006(±0.0001)
O_{eq}	3.9(±0.9)	2.40(±0.02)	0.0063(±0.0015)
C	4.9(±3.4)	3.1(±0.1)	0.011(±0.008)
O_2	4.3(±2.9)	3.4(±0.1)	0.014(±0.009)
Np2	**N**	**R (Å)**	**σ^2 (Å^2)**
O_{ax}	1.8(±0.3)	1.85(±0.01)	0.0005(±0.0001)
O_{eq}	4.4(±0.8)	2.40(±0.02)	0.0061(±0.0011)
C	3.0(±2.9)	3.2(±0.1)	0.016(±0.015)
O_2	3.8(±2.5)	3.4(±0.1)	0.015(±0.010)
Np3	**N**	**R (Å)**	**σ^2 (Å^2)**
O_{ax}	2.0(±0.3)	1.86(±0.01)	0.001 (fix)
O_{eq}	4.0(±0.8)	2.40(±0.02)	0.0078(±0.0016)
C	4 (fix)	3.1(±0.1)	0.010 (fix)
O_2	4 (fix)	3.4(±0.1)	0.010 $(= \sigma^2(C))$
Np4	**N**	**R (Å)**	**σ^2 (Å^2)**
O_{ax}	2.1(±0.3)	1.86(±0.01)	0.0011(±0.0002)
O_{eq}	4.1(±0.7)	2.41(±0.02)	0.0059(±0.0010)
C	4 (fix)	3.2(±0.1)	0.014 (fix)
O_2	4 (fix)	3.4(±0.1)	0.014 $(= \sigma^2(C))$
U1	**N**	**R (Å)**	**σ^2 (Å^2)**
O_{ax}	1.7(±1.3)	1.83(±0.01)	0.0066(±0.0050)
O_{eq}	5.6(±1.2)	2.38(±0.02)	0.0050 (fix)
C_{bi}	1.9(±1.0)	2.9(±0.1)	0.0031(±0.0016)
C_{mono}	1.9(±1.2)	3.3(±0.1)	0.0031 $(= \sigma^2(C_{bi}))$

we compare these coordination structures to our EXAFS results, we find good agreement between R(Oax)) 1.86 Å and that for the solid Np(V) phases described above (1.82 – 1.90

4. Results and discussion

Å). In contrast to the previously mentioned known structures with six–fold equatorial coordination, we find an Oeq coordination number of about four. Coordination numbers are, however, associated with high uncertainties (Table 4.11). It is known from bond valence theory that bond length and coordination numbers are correlated [96] so that bond lengths can serve as a measure for coordination number. The Np–Oeq bond length is found to be 2.4 Å. The observed coordination environment results in a bond valence sum for the central Np atom of BVS(Np) = 4.9. Assuming fivefold equatorial coordination at the same bond distance it would yield: BVS(Np) = 5.3. From this perspective the fourfold coordination is more likely, too.

The bond lengths associated with bidentate aqueous neptunyl–carbonato complexes (Np–Oeq 2.48-2.53 Å, Np–C distance 2.93–2.98 Å) have a longer Np–O distance and a shorter Np–C distance than the corresponding bond lengths obtained from analysis of our calcite samples (Np–Oeq distance 2.4 Å, Np–C 3.1-3.2 Å). This difference can only be explained as monodentate bonding of the coordinating carbonate groups to the central neptunyl ion in our samples. The equal coordination numbers for Oeq and carbon (C) are also an indication for monodentate carbonate complexation; a bidentate coordination would necessarily cause an increase in N(Oeq)/N(C). In the calcite crystal structure, Ca^{2+} ions are coordinated by six monodentate bound carbonate ions. The inter atomic distances are Ca–O1, 2.35 Å, Ca–C, 3.20 Å, and Ca–O2, 3.44 Å [37]. If we hypothesize that neptunyl occupies a calcium site in the calcite structure with only four monodentate bound carbonate ions in the first equatorial coordination sphere, the resulting two vacant carbonate sites could leave space for the axial neptunyl oxygen atoms. The Np–Oeq distance (2.4 Å) is significantly longer than the Ca–O1 distance in the calcite structure, but typical for four–fold equatorial oxygen coordination around neptunyl [124]. Np–C and Np–O2 distances are within the limits of error the same as Ca–C and Ca–O2 distances in the calcite structure. These observations imply that the calcite lattice structure slightly relaxes to fulfill the bond length requirements of the neptunyl ion.

The relaxation of the calcite structure by neptunyl incorporation suggests a positive enthalpy of mixing. Therefore, in a solid-solution series between a neptunyl–carbonate phase and calcite only limited miscibility can be expected, in particular at room temperature. The low Debye—Waller factors obtained from the EXAFS analysis for the Oax and Oeq scattering paths indicate a high structural order in the coordination polyhedron of the Np atoms. We interpret the increased Debye–Waller factors (0.010-0.016 $Å^2$), together with the relatively high uncertainties in bond lengths and coordination numbers for the C and O2 scattering paths, as indication of disorder in the orientation of the carbonate ions coordinated around the neptunyl ions. That we find solely monodentate carbonate complexation around the neptunyl ion is a strong argument for the structural incorporation of the neptunyl ions into the calcite structure during the coprecipitation experiments. The small disorder introduced into the calcite structure by incorporation of neptunyl suggests a high stability of this structure. A picture of a neptunyl ion incorporated into the calcite structure according to our results is shown in Figure 4.31.

The two missing carbonate ions and the substitution of Ca^{2+} by NpO_2^+ leave a charge

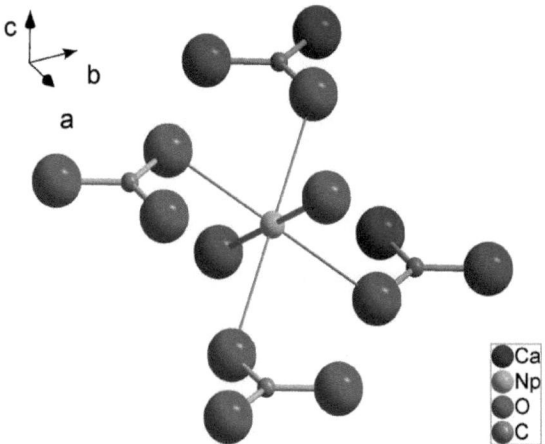

Fig. 4.31: Structural model of a neptunyl-ion incorporated into the calcite host. Neptunium(V) is located on a Ca^{2+} site and the two axial neptunyl oxygens substitute for two adjacent carbonate ions. The resulting four monodentate bound carbonate ions and six coordinating Ca^{2+} ions are shown. The coordinate system indicates a possible orientation of this structure relative to the hexagonal calcite lattice.[27]

excess of +3 in the calcite structure upon neptunyl incorporation. We do not yet have evidence how this charge is balanced. Possible charge balancing mechanisms could be a coupled substitution of Ca^{2+} ions by Na^+ or creation of vacant Ca^{2+} sites. Site–selective

4. Results and discussion

time resolved laser fluorescence investigations of the charge balancing mechanism in Eu(III) containing calcite show strong evidence for the coupled substitution of Na^+ and Eu^{3+} for 2 Ca^{2+} [82]. An analogous charge balancing mechanism with coupled substitution of Na^+ and NpO_2^+ for Ca^{2+} might also apply here. The critical EXAFS scattering path yielding information about such a coupled substitution mechanism would be the calcium shell surrounding the incorporated neptunyl. However, the distance to the Ca coordination sphere at 4.03 Å [37] is too long to be identified from room temperature EXAFS data. Attempts to fit this shell resulted in unreasonably high Debye—Waller factors.

Due to the low uranyl content of sample U1 of about 23 ppm, the usable k–range of the EXAFS spectra of this sample is limited (1.8 – 8.5 Å^{-1}). Data and model curves in k– and R–space are shown in Figure 4.30. The EXAFS analysis yields the following results: two Oax at 1.83 Å, five to six (5.6 ($\pm$1.2) Oeq at a distance of 2.38 Å from the central U atom. This distance corresponds well with the equatorial bond lengths of U(VI) phases exhibiting five–fold Oeq coordination, 2.37 ($\pm$0.09) Å [124]. For the carbon coordination sphere, the best fit result is obtained if we split the carbon scattering path into two sub-shells, resulting in carbon atoms at 2.87 Å (Cbi) and at 3.33 Å, (Cmono) (Table 4.11). This result could be explained by mixed bidentate and monodentate coordination of the carbonate ions around the uranyl in the equatorial plane.

Coordination numbers obtained for the Cbi and Cmono scattering paths (1.9 $\pm$1.0 and 1.9 $\pm$1.2) make it impossible to differentiate if there is one bidentate and three monodentate bound carbonate ions or two bidentate and only one monodentate bound carbonate in the equatorial plane around the uranyl ion. We tend to interpret this as an indication that there is likely a mixture of the two configurations. Ball and stick representations of the two possible configurations are shown in Figure 4.32.

Reeder et al. 2000 and 2001 [23, 24] find a very similar uranyl environment in synthesized uranyl doped calcite. They report that the uranyl ion is coordinated by five equatorial oxygens originating from three or four carbonate ions, two or one of them bound in bidentate fashion. The bond length to the closest equatorial oxygens Oeq, 2.33 Å, is remarkably short in their case. The simultaneous presence of bidentate and monodentate bound carbonate ions suggests that upon coprecipitation uranyl is incorporated into a strongly distorted site, which is not compatible to the calcite structure.

Kelly et al. 2003 and 2006 [25, 26] report that uranyl can occupy a stable lattice position

4. Results and discussion

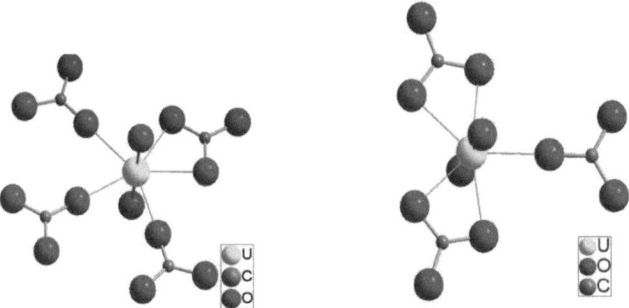

Fig. 4.32: Two equatorial coordination environments are possible for uranyl coprecipitated with calcite: one bidentate and three monodentate bound carbonate groups or two bidentate and only one monodentate bound carbonate groups.

in natural calcite samples. From their investigation of U(VI) containing natural calcite samples of different age (13.7 ka and 298 Ma), they suggest that the uranyl environment in calcite might evolve and become more calcite compatible over long time scales. The structure they suggest for the uranyl environment in the younger calcite is very similar to the neptunyl environment in our coprecipitated calcite. If indeed the structural environment of uranyl incorporated into calcite becomes more calcite compatible over geological time spans, whereas neptunyl is incorporated into such a stable site immediately upon coprecipitation, then this is another argument for the higher compatibility of neptunyl and calcite compared to uranyl and calcite. It is, however, difficult to draw conclusions from the comparison of natural and synthesized Np and U doped calcite, as the conditions during formation (e.g., supersaturation, temperature, and uranyl activity) and during geologic periods of time (e.g., diagenesis, heat, and pressure) are usually not precisely known for natural samples.

Raman and NIR spectroscopy

The highly doped calcite from experiments Np9 - Np10 (12000 ppm) has been used for a Raman spectroscopic investigation of the neptunyl(V) incorporation species. A spectrum of the neptunyl doped calcite (red) in comparison with pure calcite (black) is shown in Figure 4.33. The observed peaks can be assigned to vibrational modes of the calcite lat-

4. Results and discussion

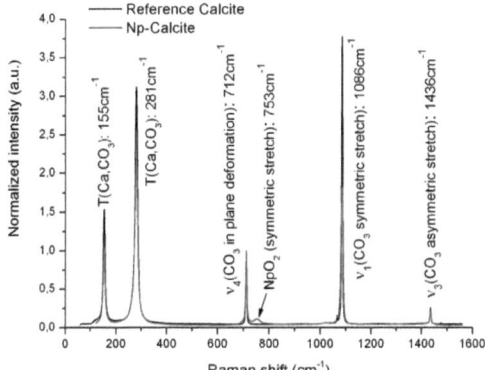

Fig. 4.33: Raman spectra of neptunyl doped and pure calcite.

tice, the carbonate molecules [125], and the neptunyl molecule. Peaks at 155 cm^{-1} and 281 cm^{-1} belong to translational lattice modes (T(Ca, CO$_3$) in Figure 4.33). The peak at 712 cm^{-1} is assigned to the ν_4– symmetric in–plane CO$_3$ deformation, the peak at 1086 cm^{-1} is related to the ν_1–symmetric CO$_3$ stretching, and the one at 1436 cm^{-1} to the ν_3–asymmetric CO$_3$ stretching. No shift in the calcite peaks is observed upon neptunyl incorporation. At much higher neptunyl loadings (on the order of magnitude of 10 %) shifts might be expected.

The main difference between the two spectra is the additional band observed at the neptunyl doped calcite at 753 cm^{-1} that is related to the ν_1–symmetric stretching vibration of the linear NpO$_2^+$ molecule. The ν_1–symmetric stretching vibration of NpO$_2^+$ is known to be sensitive to equatorial carbonate coordination. Its frequency for the pure NpO$_2^+$ aquo–ion, 767 cm^{-1}, decreases with increasing carbonate complexation: 762 cm^{-1} for NpO$_2$(CO$_3$)$^-$, 756 cm^{-1} for NpO$_2$(CO$_3$)$_2^{3-}$, and 756 cm^{-1} for NpO$_2$(CO$_3$)$_3^{5-}$ [112]. The vibrational frequency is inversely correlated to the Np–O$_{ax}$ distance which is 1.82 Å for the NpO$_2^+$ aquo–ion [126], 1.84 Å for NpO$_2$(CO$_3$)$^-$, 1.85 Å for NpO$_2$(CO$_3$)$_2^{3-}$, and 1.86 Å for NpO$_2$(CO$_3$)$_3^{5-}$ [113]. The calcite incorporated NpO$_2^+$ species fits well into this series with its Raman frequency of 753 cm^{-1} and the Np–O$_{ax}$ distance of 1.86 Å.

4. Results and discussion

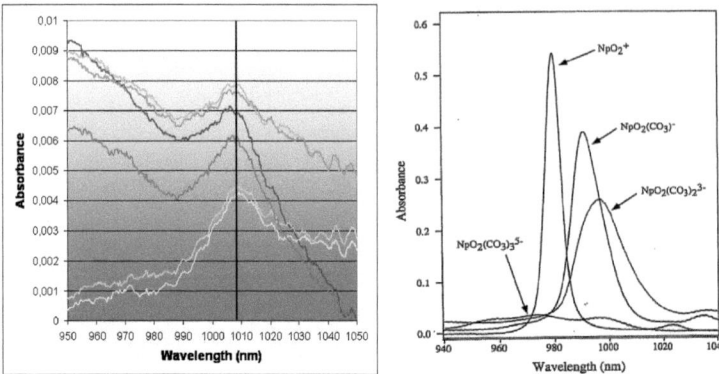

Fig. 4.34: Left diagram: NIR absorption spectra measured on three calcite single crystals from experiment Np11(SC). Right diagram: NIR absorption spectra of the neptunyl aquo–ion and the carbonato complexes [112].

A similar trend can be observed in the red shift of the NIR absorption band of the NpO_2^+ molecule with subsequent carbonate complexation. NIR absorption spectra measured on three neptunyl doped calcite single crystals and those of the neptunyl aquo–ion and the carbonato complexes [112] are shown in Figure 4.34. The peak maximum of the absorption band is at 980 nm for the NpO_2^+ aquo–ion, at about 990 nm for $NpO_2(CO_3)^-$, and at about 995 nm for $NpO_2(CO_3)_2^{3-}$. In the centro–symmetric neptunyl triscarbonato complex (cp. Figure 4.25) the f → f transition is forbidden and no absorption can be observed. In NIR spectra measured on neptunyl doped calcite single crystals from MFR experiment Np11(SC), an absorption band has been observed at 1008 nm. This agrees well with the subsequent red shift upon carbonate complexation and shows that the coordination environment of NpO_2^+ incorporated into calcite is not a perfectly centro–symmetric square–bipyramid. The molar extinction coefficient of the incorporation species could be estimated to be about 60 $M^{-1}cm^{-1}$, compared to 390 $M^{-1}cm^{-1}$ for the NpO_2^+ aquo–ion [111].

Neptunyl species, vibrational frequencies observed by Raman spectroscopy, Np–O_{ax} bond length from EXAFS investigations, and wavelengths of NIR absorption bands are summarized in Table 4.12. The spectroscopic characteristics of the calcite incorporated neptunyl species fits well into the series of the neptunyl carbonate complexes. With subsequent

4. Results and discussion

Tab. 4.12: The shift of spectroscopic parameters with subsequent carbonate complexation and calcite incorporation of NpO_2^+.

Species	Raman frequency (cm^{-1})	$R(Np\text{--}O_{ax})$ (Å)	NIR absorption band (nm)
NpO_2^+ aquo–ion	767 [112]	1.82 [126]	980 [112]
$NpO_2(CO_3)^-$	762 [112]	1.84 [113]	$\sim$ 990 [112]
$NpO_2(CO_3)_2^{3-}$	756 [112]	1.85 [113]	$\sim$ 995 [112]
$NpO_2(CO_3)_3^{5-}$	756 [112]	1.86 [113]	no absorption (point symmetry)
NpO_2^+–calcite incorporation species	753	1.86 [27]	1008

carbonate coordination the electron density at the neptunium center increases, thereby increasing the Np–O_{ax} bond length and subsequent reduction in vibrational frequency of the symmetric stretching vibration. At the same time the energy needed to excite the f–electrons decreases. The spectroscopic characteristics indicate relatively strong bonds between the four monodentate carbonate ions in the equatorial coordination of the calcite incorporated neptunyl, comparable to the six bonds in the neptunyl triscarbonato complex.

4.3.3. Solid solution thermodynamics for neptunyl(V) doped calcite

It is important to emphasize that the D values derived from MFR experiments describe the composition of a mixed phase precipitated from a supersaturated solution of a defined composition. As mentioned in section 2.5.1, previous site specific incorporation studies indicate that incorporation of foreign ions into calcite depends on crystal nano–topography (non–equivalent behavior of acute and obtuse steps causing intrasectoral zoning), indicating non–equilibrium conditions [22, 47]. Similar effects have to be expected for neptunyl coprecipitation with calcite.

According to these arguments it is not suitable to describe the system by equilibrium thermodynamics. However, as mentioned already in section 2.5.1, there is hardly any other way to investigate structural incorporation at room temperature from aqueous environ-

4. Results and discussion

ments. Therefore it is common practice to perform coprecipitation experiments at low supersaturation and low growth rates and to interpret the results in terms of equilibrium thermodynamics. In MFR experiments presented here steady state SI varies between 0.04 and 0.95. Solutions are metastable with respect to homogeneous calcite nucleation within the time frame of the experiments [67]. Crystal growth is surface controlled and growth rates are low compared to other coprecipitation studies [122].

According to these arguments equilibrium thermodynamics might be applicable to the system even though it is not strictly correct. Another topic that complicates the application of equilibrium solid solution thermodynamics to neptunyl coprecipitation with calcite is that the exact substitution and charge compensation mechanism is not known. In other words, the neptunium endmember of the mixing series is unknown. However, likely assumptions about the charge compensation can be made according to the results of the EXAFS analysis. The EXAFS results indicate that NpO_2^+ substitutes Ca^{2+} and two carbonate ions in the calcite lattice. Let us assume that charge compensation is provided by additional vacant Ca^{2+} sites. The overall substitution mechanism then becomes:

$$Ca_3(CO_3)_3 \leftrightarrow Ca_{(0.5)}NpO_2CO_3 + 1.5 \text{ vacant } Ca^{2+} \text{ sites}$$

This corresponds to a solid solution composition of:

$$Ca_{(3-2.5X)}(NpO_2)_X(CO_3)_{(3-2X)} \; ; \qquad \text{with } 0 \leq X \leq 1$$

For MFR experiments presented here the mole fraction, X, ranges from 0 to 0.036. Due to the limited experimental range, an optimal description of the experiments can be achieved using the simple model of a regular solid solution. Simpler models, e.g., using a constant solid solution activity coefficient, do not describe the data well. More complicated models, e.g., assuming a subregular solid solution resulted in insignificant parameters. Equations needed for the calculation are derived in section 2.5.1. The calculation starts from equation 2.15:

$$D = \frac{K_B \gamma_A}{K_A \gamma_B} \exp(a_0(X_B^2 - X_A^2)).$$

K_A is in this case the solubility product of calcite to the power of three, K_{SP}^3(Calcite). K_B is the solubility product of a hypothetical neptunium endmember phase, K_{SP}(Np), $X_B = X$ and $X_A = 1 - X$. γ_A is the product of the aqueous activity coefficients according to the ion concentration product of $Ca_3(CO_3)_3$, for now called γ_{Ca}, and γ_B is the

4. Results and discussion

corresponding product for $Ca_{(0.5)}NpO_2CO_3$, for now called γ_{Np}. Note that the D values needed for this calculations are not equivalent to the empirical D values listed in Table 4.10. Thermodynamic partition coefficients must be calculated according to equation 2.6. Introducing these parameters we get:

$$D = \frac{K_{SP}(\text{Np})\,\gamma_{Ca}}{K_{SP}^3(\text{Calcite})\,\gamma_{Np}} \exp(a_0(2X - 1)). \tag{4.1}$$

Taking the natural logarithm of the equation we obtain the equation of a straight line:

$$\ln D = \left(\ln\left(\frac{K_{SP}(\text{Np})\,\gamma_{Ca}}{K_{SP}^3(\text{Calcite})\,\gamma_{Np}}\right) - a_0\right) \cdot 2a_0 X \tag{4.2}$$

The unknown parameters a_0 and $K_{SP}(\text{Np})$ can now be derived from a linear regression of $\ln D$ versus X. This is demonstrated for the MFR experiments of this study at pH about 10.2 in Figure 4.35. The parameters calculated for the calcite–neptunyl solid solution are

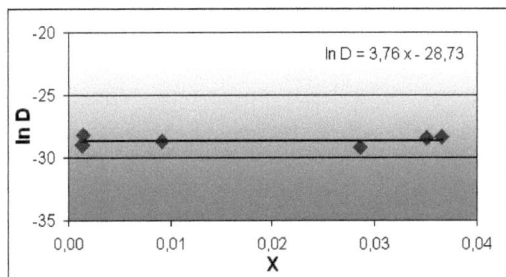

Fig. 4.35: Solid solution thermodynamic parameters can be derived from linear regression of $\ln D$ versus mole fraction X.

$a_0 \approx 2 \pm 5$ and $\log_{10}(K_{SP}(\text{Np})) \approx -13 \pm 2$. A Guggenheim parameter a_0 of about 2 indicates a positive excess free energy of mixing. The related solid solution activity coefficients range from 6.7 to 1. The solubility product of the hypothetical neptunium endmember should theoretically be larger than the the solubility product of an existing phase with the same stoichiometry. A phase with composition $Ca_{(0.5)}NpO_2CO_3$ has been reported [40], but no solubility data is available. The most similar phase solubility data is available for is $NaNpO_2CO_3$ ($\log_{10} K_{SP} = -11.66$ [32]). This value is higher, but due to the different compositions the values are not directly comparable.

4. Results and discussion

Even though the parameters seem plausible they have large uncertainties and their derivation is based on many assumptions. Therefore, further applications that can be calculated with these parameters such as the Lippmann–Diagram (cp. section 2.5.1) are not shown here. The parameters should not be considered for any kind of predictive modeling. The calculation just shows how in principle thermodynamic parameters can be derived quite simply from MFR experiments even for solid solutions involving complex substitution mechanisms.

4.4. Comparison of neptunyl and uranyl interactions with calcite

The results from MFR experiments clearly show that neptunyl has a higher affinity for incorporation into calcite than uranyl. EXAFS analyses revealed equatorial coordination by four monodentate bound carbonate ions for calcite incorporated neptunyl ions. This suggests that neptunyl is structurally incorporated into the calcite host upon coprecipitation. Neptunyl substitutes one Ca^{2+} and two carbonate ions in the calcite structure.

EXAFS results reported in literature [23, 24] and obtained in this study for uranyl coprecipitation with calcite show that upon coprecipitation uranyl is incorporated onto a highly distorted site. It keeps one or two bidentate bound carbonate ions in its equatorial coordination. This structure is not compatible with the calcite structure.

Investigation of natural uranyl containing calcite samples show that over geologic time spans the coordination environment of uranyl in calcite adepts to the calcite structure [25, 26] and becomes similar to the one observed for neptunyl coprecipitated with calcite. The different structural compatibility of uranyl and neptunyl with calcite results in empirical partition coefficients for uranyl and neptunyl coprecipitation with calcite differing by one to two orders of magnitude, $D(U(VI)) = 0.02$; $D(Np(V)) = 0.2 - 10.3$. This likely reflects the higher affinity of uranyl for bidentate carbonate complexation compared to neptunyl, that is also indicated by the higher formation constants for aqueous uranyl carbonato complexes compared to neptunyl carbonato complexes.

Neptunyl adsorbs at the calcite surface preferredly at step edges as a bidentate inner–sphere complex (cp. section 4.2.3). During coprecipitation experiments adsorption at step sites blocks these active surface growth sites and causes a decrease in calcite crystal

4. Results and discussion

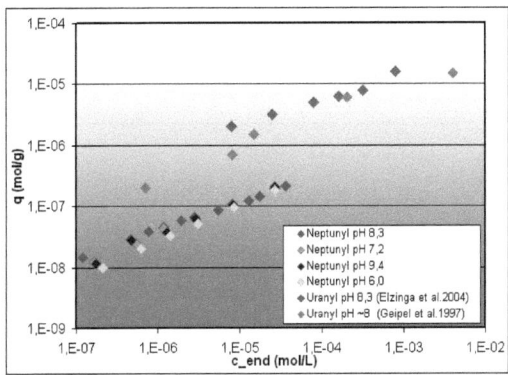

Fig. 4.36: Neptunyl adsorption isotherms from this study compared to uranyl adsorption data from literature [80, 21]

growth rate (cp. Figure 4.29). Uranyl has been reported to adsorb at the calcite surface as a triscarbonato complex [21, 80]. Figure 4.36 shows that under equal conditions more uranyl cations adsorb at the calcite surface than neptunyl. KD values for uranyl adsorption onto calcite are about one order of magnitude higher than for neptunyl. Nevertheless, no growth retardation effect has been observed in the uranyl coprecipitation experiment (cp. Figure 4.29), which suggests that uranyl adsorbs at the calcite surface mainly in an outer–sphere fashion. Only a very limited amount of uranyl may adsorb at step edges, as proposed by Reeder et al. (2004) [22], where it becomes captured by the growing calcite crystal and incorporated.

Chapter 5

Conclusions

In summary the most important experimental results obtained during this thesis work are:

- Calcite zetapotential

    PALS zetapotential measurements show the expected shift of the calcite IEP with changing CO_2 partial pressure.

    Streaming potential measurements reveal the importance of Ca^{2+} and CO_3^{2-} for the zetapotential and show a limited influence of pH.

- The calcite(104)–water interface structure

    The part of the calcite(104)–water interface structure surface diffraction measurements probe does not change significantly, even upon the extreme changes in the composition of the contact solution investigated.

    No indication for calcium or carbonate inner–sphere complexes on the flat calcite (104)–face has been observed.

    Two well ordered layers of water at 2.35 ± 0.05 Å and 3.24 ± 0.06 Å above the calcite surface have been identified. Water molecules of the first layer are located above the surface calcium ions and those of the second layer are located above the surface carbonate ions.

    In contact to solution mainly carbonate ions at the calcite surface relax from their bulk position and tilt towards the surface by about 4°. The other ions in the first two calcite monolayers show only slight relaxation. A subsequent increase in Debye–Waller factors is observed from the bulk to the surface.

5. Conclusions

- Surface complexation modeling

    A new Basic Stern SCM is presented that describes the measured zetapotentials. According to the observations made with surface diffraction only outer–sphere complexes of ions other than protons and hydroxide are considered.

    It is suggested that inner–sphere adsorption may take place at step and kink sites on the calcite surface, sites that are known to be most reactive during crystal growth and dissolution. Processes taking place at these sites could explain the effects observed during zetapotential measurements that are not described by the SCM. Surface diffraction measurements are not sensitive for structural changes at step sites.

- Adsorption of neptunyl(V) at calcite

    Adsorption has been quantified over a large range of pH and concentration conditions. Adsorption of neptunyl at calcite shows a Freundlich–like concentration dependence. Adsorption is maximal at pH 8.3 and decreases with increasing and decreasing pH. The pH dependence of adsorption is greater at low concentrations; above a equilibrium concentration of 10^{-5} M NpO_2^+ no significant pH dependence is observed. Maximum surface loading corresponds to 2 % surface site coverage.

    Compared to uranyl(VI), the KD values for neptunyl(V) adsorption at calcite are about one order of magnitude lower.

    EXAFS analyses reveals that the most abundant adsorption complex at pH 8.3 and high surface coverage is most likely a bidentate inner–sphere biscarbonato complex bound at step edge sites on the calcite (104)–face.

    Adsorption kinetic experiments and desorption experiments indicate that a small part of adsorbed neptunyl is likely incorporated into calcite, even at calcite equilibrium conditions. Spectroscopic evidence for this incorporation processes has not been found.

- Neptunyl(V) coprecipitation with calcite

    Upon coprecipitation at surface controlled calcite growth conditions, neptunyl(V) is readily incorporated into the calcite structure. It most likely substitutes a calcium and two adjacent carbonate ions in the calcite structure. Neptunyl(V) incorporated

5. Conclusions

into the calcite structure shows, therefore, an equatorial coordination by four monodentate carbonate ions.

Neptunyl(V) shows a much higher affinity for incorporation into calcite than uranyl(VI). Empirical partition coefficients for neptunyl coprecipitation with calcite range from 0.2 to 10.3. For uranyl a value of 0.02 is observed. This could be related to the higher affinity of uranyl for the formation of bidentate carbonate complexes. Uranyl is observed to retainits bidentate carbonate ligands in the equatorial coordination sphere even upon coprecipitation. The associated equatorial environments are not compatible with the calcite structure.

Reduced calcite growth rates are observed in neptunyl coprecipitation experiments, compared to pure calcite precipitation experiments. Effective growth retardation is expected for ions that adsorb at step and kink sites, the most active sites during crystal growth. This agrees well with the adsorption complex structure identified from EXAFS investigation of the adsorption sample.

No crystal growth retardation effects are observed for uranyl coprecipitation with calcite. As more uranyl adsorbs at calcite compared to neptunyl, this can only be explained if uranyl is assumed to adsorb at the calcite surface mainly as an outer–sphere complex.

The comparison of uranyl(VI) and neptunyl(V) regarding the interactions with calcite is intriguing. The actinyl cations are as a first approximation often expected to behave similarly. Their differing interactions with calcite demonstrate how different the chemistry of neptunyl(V) and uranyl(VI) in fact is and how erroneous the assumption of equal chemical behavior may be.

The differences in adsorption, coprecipitation, and growth retardation behavior are a good example of how important it is to study processes at a molecular scale. Reliable explanation of the differences in the observed sorption behavior is only possible taking into account the molecular structure of solution, adsorption, and incorporation species.

As this study has been performed in the context of nuclear waste disposal, it is an important concluding remark to answer the question: "What are the consequences of the observed neptunyl(V) calcite interactions for the long term safety of a nuclear waste repository?" As explained in the introduction, calcite is expected to be present in most of the

5. Conclusions

host rock formations considered for nuclear waste repositories as well in the near and far fields around the nuclear waste disposal sites.

Adsorption of neptunyl(V) at the calcite surface will surely immobilize some of the neptunyl(V) dissolved in subsurface water and thereby retard its migration through the geosphere. Surface adsorption is an effective retardation mechanism, but it is highly reversible. It has been shown that neptunyl is readily incorporated into the calcite structure upon calcite precipitation. However, as calcite precipitation takes place only in very special geochemical milieus, this process is expected to be relevant only under special circumstances. A highly relevant retardation mechanism for the longterm fate of neptunyl(V) in the geosphere is the structural incorporation into calcite at calcite equilibrium conditions. Such reactions are to some extent expected if the resulting solid solution is thermodynamically stable. Indications that such processes might take place have been found in batch type adsorption kinetics and desorption experiments. Indications for the thermodynamic stability of neptunyl doped calcite could be derived from the compatibility of the structure of the incorporation species with the calcite crystal structure and from the estimation of thermodynamic parameters based on MFR experiments.

The "European Pilot Study on the Regulatory Review of the Safety Case for Geological Disposal of Radioactive Waste" implicitly identifies processes understanding as one of the key issues in reducing the uncertainties related to nuclear waste disposal [127]. The process understanding gained on the adsorption of neptunyl(V) onto calcite surfaces and on the incorporation of neptunyl(V) into calcite can be used as a supporting argument in the safety case for a nuclear waste repository. It reduces the uncertainty related to the expected mobility of neptunyl(V) in groundwater aquifers. In order to enable more substantiated statements about neptunyl(V) mobility, an important next step will be to derive thermodynamic and/or kinetic parameters for the observed reactions that can be used in geochemical transport models.

Calcite is an potentially important sink for neptunyl as well as for many other heavy metal ground water contaminants. However, the real nature of its surface reactivity, how exactly adsorption and incorporation reactions proceed, and to which extent thermodynamic parameters for these reactions can be derived, on the one hand, and how reasonable it will be to apply them, on the other hand, leaves still space for many open questions.

Chapter 6

Acknowledgment

This doctoral work has been performed between October 2006 and November 2009 at the Institute for Nuclear Waste Disposal (INE) at the Karlsruhe Institute of Technology (KIT) (former Forschungszentrum Karlsruhe). I would like to acknowledge all the INE staff members for their helpful contributions to the success of this thesis and for the amiable and productive atmosphere at the institute.

I especially would like to acknowledge my thesis supervisor Professor Dirk Bosbach for encouraging comments and interesting discussions and his way of mentoring that guided me to independent scientific work.

Special thanks goes to Professor Thomas P. Trainor who kindly agreed to be second reviewer of this thesis. His support during the analysis of CTR data and his hospitality during my stay at the University of Alaska Fairbanks are gratefully acknowledged.

I sincerely thank Professor Thomas Neumann from the institute of mineralogy and geochemistry that he agreed to be the third reviewer of this thesis.

I would like to thank Professor Horst Geckeis and Professor Thomas Fanghänel current and former heads of the Institute for Nuclear Waste Disposal (INE) for allowing me to perform my doctoral research at the INE and providing me with constructive critics and expert advice.

My special thanks goes to Dr. Melissa A. Denecke who introduced me to synchrotron work, for her help and advice during EXAFS measurements and data analysis, and for supporting me and providing me inhouse beam–time to experiment at the INE–beamline. I'm very grateful that she helped me get started with CTR work at the APS. Her thorough review of this manuscript was of invaluable help and is gratefully acknowledged.

6. Acknowledgment

I sincerely thank Dr. Johannes Lützenkirchen who taught me most of what I know about surface complexation modeling. His comments led to significant improvements of this research work. His thorough review of this manuscript was of great help.

I would like to express my special thanks to Dr. Boris Brendebach, Dr. Kathy Dardenne, Dr. Jörg Rothe, and Dr. Tonja Vitova, staff members of the INE–beamline, for their expert advice and help concerning all around EXAFS.

I especially acknowledge outstanding support, expert advice, and hospitality of Dr. Peter Eng, contact at GeoSoilEnviroCARS (Sector 13), Advanced Photon Source (APS), Argonne National Laboratory, where parts of this work were conducted. GeoSoilEnviroCARS is supported by the National Science Foundation - Earth Sciences (EAR-0622171) and Department of Energy - Geosciences (DE-FG02-94ER14466). Use of the Advanced Photon Source was supported by the U. S. Department of Energy, Office of Science, Office of Basic Energy Sciences, under Contract No. DE-AC02-06CH11357.

I want to thank Dr. Andreas C. Scheinost very much for his support during low temperature EXAFS measurements at the Rossendorf beamline. The provision of beamtime at the Rossendorf beamline (ROBL) by the European Synchrotron Facility in Grenoble is gratefully acknowledged.

Special recognition goes to Dr. Patrick Lindqvist–Reis for his help with the Raman spectroscopic investigations and his helpful comments on actinide spectroscopy.

For her kind introduction to CTR data analysis and her memorable hospitality I sincerely acknowledge Dr. Sarah Petitto.

I want to thank Dr. Christian Marquardt for his help on working in the controlled area laboratories and Dr. Nidhu Lal Banik for the preparation of the neptunyl(V) stock solution. I thank Dr. Dieter Schild and Dr. Andreas Bauer for XPS and powder–XRD analysis of calcite samples.

For help in the laboratory, ICP–MS–, ICP–OES–, and N_2–BET–analyses I want to thank Stephanie Heck, Elke Bohnert, Cornelia Walschburger, Silvia Rabung, and Tanja Kisely.

For SEM analyses and calcite sample preparation thanks go to Eva Soballa.

The INE radio protection crew namely: Jens Tomas, Elfi Nagel, and Gerhard Cristill, is acknowledged for help and support on radioactive sample transport.

Most experiments would not have been possible without especially designed equipment produced by the INE workshop. In this context I would like to thank Günther Hermann, Volker Krepper, and Erwin Schmitt.

6. Acknowledgment

I want to thank all the guys from room 207c for their helpfullness and the nice atmosphere. Special thanks to Holger Seher, Eva Hartmann, Moritz Schmidt, Christoph Burmeister, and Marie Marques–Fernandez.

Generous financial support by the Karlsruhe House of Young Scientists that allowed me to stay in the US for two months and to invite Professor Thomas P. Trainor to my disputation is gratefully acknowledged.

This work has been part of the european integrated project FUNMIG, the Helmoltz Virtual Institute: Advanced Solid–Aqueous RadioGeochemistry, and RECAWA, a project in the DFG R&D program "Geotechnologien". The financial support is gratefully acknowledged.

Bibliography

[1] Barthelmy, D. http://webmineral.com (1997-2009).

[2] Kühnle, J. http://www.wissen-im-netz.info/mineral/lex/abc/c/calcit.htm (1999–2007).

[3] Taubald, H. *et al.* Experimental investigation of the effect of high-pH solutions on the opalinus shale and hammerschmiede smectite. *Clay Minerals* **35**, 515–524 (2000).

[4] Carlsson, T. & Aalto, H. Coprecipitation of Ni with calcite: An experimental study. In *Scientific Basis for Nuclear Waste Management XXI*, vol. 506 of *Materials Research Society Symposium Proceedimgs*, 621–627 (Materials research society, 1998).

[5] Rimstidt, J. D., Balog, A. & Webb, J. Distribution of trace elements between carbonate minerals and aqueous solution. *Geochimica et Cosmochimica Acta* **62**, 1851–1863 (1998).

[6] Curti, E. Coprecipitation of radionuclides: basic concepts, literature review and first applications. In *PSI-Bericht*, vol. 97 (PSI, 1997).

[7] Zachara, J. M., Cowan, C. E. & Resch, C. T. Sorption of divalent metals on calcite. *Geochimica et Cosmochimica Acta* **55**, 1549 – 1562 (1991).

[8] Gompper, K. Zur Abtrennung langlebiger Radionuklide. In *Radioaktivität und Kernenergie*, chap. 10, 153–167 (Forschungszentrum Karlsruhe GmbH, 2000), 2 edn.

[9] Katz, J. J., Seaborg, G. T. & Morss, L. R. *The chemistry of the actinide elements*, vol. 1 (Chapman and Hall, London, New York, 1986), 2 edn.

Bibliography

[10] Pfennig, G., Klewe-Nebenius, H. & Seelmann-Eggebert, W. *Karlsruher Nuklidkarte* (Marktdienste Haberbeck GmbH, Lage, Lippe, 1998), 6 edn.

[11] Dozol, M. & Hagemann, R. Radionuclide migration in groundwaters: Review of the behaviour of actinides. *Pure & Applied Chemistry* **65**, 1081–1102 (1993).

[12] Clark, D. L., Hobart, D. E. & Neu, M. P. Actinide carbonate complexes and their importance in actinide environmental chemistry. *Chemical Reviews* **95**, 25–48 (1995).

[13] Antonio, M. R., Soderholm, L., Williams, C. W., Blaudeau, J. P. & Bursten, B. E. Neptunium redox speciation. *Radiochimica Acta* **89**, 17–25 (2001).

[14] Denecke, M. A. Actinide speciation using x-ray absorption fine structure spectroscopy. *Coordination Chemistry Review* **250**, 730–754 (2006).

[15] Zhong, S. & Mucci, A. Partitioning of rare earth elements between calcite and seawater solutions at 25°C and 1 atm, and high dissolved REE concentrations. *Geochimica et Cosmochimica Acta* **59**, 443–453 (1995).

[16] Elzinga, E. *et al.* EXAFS study of rare-earth element coordination in calcite. *Geochimica et Cosmochimica Acta* **66**, 2875–2885 (2002).

[17] Curti, E., Kulik, D. A. & Titts, J. Solid solutions of trace Eu(III) in calcite: Thermodynamic evaluation of experimental data over a wide range of pH and p(CO_2). *Geochimica et Cosmochimica Acta* **69**, 1721–1737 (2005).

[18] do Sameiro Marques Fernandes, M. *Spektroskopische Untersuchungen (TRLFS und XAFS) zur Wechselwirkung von dreiwertigen Lanthaniden und Actiniden mit der Mineralphase Calcit*. Ph.D. thesis, Institut für Nukleare Entsorgung / Forschungszentrum Karlsruhe (2006).

[19] Stumpf, T. & Fanghänel, T. A time resolved laser fluorescence spectroscopy (TRLFS) study of the interaction of trivalent actinides (Cm(III)) with calcite. *Journal of Colloid and Interface Science* **249**, 119–122 (2002).

[20] Sturchio, N., Antonio, M., Soderholm, L., Sutton, S. & Brannon, J. Tetravaalent uranium in calcite. *Science* **281**, 971–973 (1998).

Bibliography

[21] Elzinga, E. J. et al. Spectroscopic investigation of U(VI) sorption at the calcite-water interface. *Geochimica et Cosmochimica Acta* **68**, 2437–2448 (2004).

[22] Reeder, R. J. et al. Site-specific incorporation of uranyl species at the calcite surface. *Geochimica et Cosmochimica Acta* **68**, 4799–4808 (2004).

[23] Reeder, R. J., Nugent, M., Lamble, G. M., Tait, C. D. & Morris, D. E. Uranyl incorporation into calcite and aragonite: XAFS and luminescence studies. *Environmental Science & Technology* **34**, 638–644 (2000).

[24] Reeder, R. J. et al. Coprecipitation of uranium(VI) with calcite: XAFS, micro-XAS and luminiscence characterization. *Geochimica et Cosmochimica Acta* **65**, 3491–3503 (2001).

[25] Kelly, S. et al. Uranyl incorporation in natural calcite. *Environmental Science & Technology* **37**, 1284–1287 (2003).

[26] Kelly, S., Rasbury, E. T., Chattopadhyay, S., Kropf, A. J. & Kremner, K. Evidence of a stable uranyl site in ancient organic rich calcite. *Environmental Science & Technology* **40**, 2262–2268 (2006).

[27] Heberling, F., Denecke, M. A. & Bosbach, D. Neptunium(V) coprecipitation with calcite. *Environmental Science & Technology* **42**, 471–476 (2008).

[28] Zavarin, M., Roberts, S., Hakem, N., Sawvel, A. & Kersting, A. Eu(III), Sm(III), Np(V), Pu(V), and Pu(VI) sorption to calcite. *Radiochimica Acta* **93**, 93–102 (2005).

[29] Keeney-Kennicutt, W. L. & Morse, J. W. The interaction of $Np(V)O_2^+$ with common mineral surfaces in dilute aqueous solutions and seawater. *Marine Chemistry* **15**, 133–150 (1984).

[30] Altmaier, M., Neck, V., Lützenkirchen, J. & Fanghänel, T. Solubility of plutonium in $MgCl_2$ and $CaCl_2$ solutions in contact with metallic iron. *Radiochimica Acta* **97**, 187–192 (2009).

[31] Dong, W. & Brooks, S. C. Determination of the formation constants of ternary complexes of uranyl and carbonate with alkaline earth metals (Mg^{2+}, Ca^{2+}, Sr^{2+},

and Ba^{2+}) using anion exchange method. *Environmental Science & Technology* **40**, 4689–4695 (2006).

[32] Hummel, W., Berner, U., Curti, E., Pearson, F. & Thoenen, T. Nagra/PSI chemical thermodynamic data base 01/01. *Radiochimica Acta* **90**, 805–813 (2002).

[33] Parkhurst, D. L. & Appelo, C. A. J. User's guide to PhreeqC (version 2). Tech. Rep., US Geological Survey (1999).

[34] Keizer, M. & van Riemsdijk, W. ECOSAT. Tech. Rep., Wageningen University (1999).

[35] Mullin, J. *Crystallization* (Butterworth Heinemann, Oxford, Boston, 2001), 4 edn.

[36] Nielsen, A. E. & Toft, J. M. Electrolyte crystal growth kinetics. *Journal of Crystalgrowth* **67**, 278–288 (1984).

[37] Markgraf, S. A. & Reeder, R. J. High-temperature structure refinement of calcite and magnesite. *American Mineralogist* **70**, 590–600 (1985).

[38] Reuss, M. *Beeinflussung der feinkristallinen Ausscheidung von Calciumcarbonat durch Verfahrensparameter und die Dotierung mit ausgewählten Chloriden der Seltenen Erden*. Ph.D. thesis, Mathematisch-Naturwissenschaftliche Fakultät / Universität Köln (2003).

[39] Guillaumont, R. *et al*. *Update on the chemical thermodynamics of Uranium, Neptunium, Plutonium, Americium and Technetium*, vol. 5 of *Chemical Thermodynamics* (NEA/OECD, 2003), 1 edn.

[40] Volkov, Y. F., Tomilin, S. V., Visyashcheva, G. I. & Kapshukow, I. I. Carbonate compounds of pentavalent actinoids with alkali-metal cations. VI. x-ray structure analysis of LiNpO$_2$CO$_3$ and NaNpO$_2$CO$_3$. *Radiokhimiya* **21**, 668–672 (1979).

[41] Hartman, P. & Perdok, W. G. On the relations between structure and morphology of crystals. I. *Acta Crystallographica* **8**, 49–52 (1955).

[42] Paquette, J. & Reeder, R. J. Relationship between surface-structure, growth-mechanism, and trace-element incorporation in calcite. *Geochimica et Cosmochimica Acta* **59**, 735–749 (1995).

Bibliography

[43] Teng, H. H., Dove, P. M. & DeYoreo, J. J. Reversed calcite morphologies induced by microscopic growth kinetics: Insight into biomineralization. *Geochimica et Cosmochimica Acta* **63**, 2507–2512 (1999).

[44] Teng, H. H., Dove, P. M., Orme, C. A. & De Yoreo, J. J. Thermodynamics of calcite growth: Baseline for understanding biomineral formation. *Science* **282**, 724–727 (1998).

[45] Teng, H. H., Dove, P. M. & De Yoreo, J. J. Kinetics of calcite growth: Surface processes and relationships to macroscopic rate laws. *Geochimica et Cosmochimica Acta* **64**, 2255–2266 (2000).

[46] Burton, W. K., Cabrera, N. & Frank, F. C. The growth of crystals and the equilibrium structure of their surfaces. *Philosophical Transactions of the Royal Society of London. Series A, Mathematical and Physical Sciences (1934-1990)* **243**, 299–358 (1951).

[47] Wasylenki, L. E., Dove, P. M., Wilson, D. S. & De Yoreo, J. J. Nanoscale effects of strontium on calcite growth: An in situ AFM study in the absence of vital effects. *Geochimica et Cosmochimica Acta* **69**, 3017–3027 (2005).

[48] Fenter, P. *et al.* Surface speciation of calcite observed in situ by high-resolution x-ray reflectivity. *Geochimica et Cosmochimica Acta* **64**, 433–440 (2000).

[49] Geissbuhler, P. *et al.* Three-dimensional structure of the calcite-water interface by surface x-ray scattering. *Surface Science* **573**, 191–203 (2004).

[50] Magdans, U., Torrelles, X., Angermund, K., Gies, H. & Rius, J. Crystalline order of a water glycine film coadsorbed on the (104) calcite surface. *Langmuir* **23**, 4999–5004 (2007).

[51] Magdans, U. *Mechanismen der Biomineralisation von Calcit am Beispiel von Seeigel Stacheln: Untersuchung der Wechselwirkung zwischen Sorbat-Molekulen und Calcit-Wachstumsgrenzflächen mit Oberflächen-Röntgenbeugung und numerischen Simulationen*. Ph.D. thesis, Institut für Mineralogie und Kristallographie, Ruhr-Universität Bonn (2005).

Bibliography

[52] Aquilano, D., Calleri, M., Natoli, E., Rubbo, M. & Sgualdino, G. The (104) cleavage rhombohedron of calcite: theoretical equilibrium properties. *Materials Chemistry and Physics* **66**, 159–163 (2000).

[53] Spagnoli, D., Kerisit, S. & Parker, S. C. Atomistic simulation of the free energies of dissolution of ions from flat and stepped calcite surfaces. *Journal of Crystal Growth* **294**, 103–110 (2006).

[54] Spagnoli, D., Kerisit, S. & Parker, S. C. Atomistic simulation of the free energies of dissolution of ions from flat and stepped calcite surfaces. *Journal of Crystal Growth* **294**, 103–110 (2006).

[55] Kerisit, S. & Parker, S. Free energy of adsorption of water and metal ions on the (10-14) calcite surface. *Journal of the American Chemical Society* **126**, 10152–10161 (2004).

[56] Kerisit, S. & Parker, S. Free energy of adsorption of water and calcium on the (104) calcite surface. *Chemical Communications* **1**, 52–53 (2004).

[57] Villegas-Jimenez, A., Mucci, A. & Whitehead, M. A. Theoretical insights into the hydrated (104) calcite surface: Structure, energetics, and bonding relationships. *Langmuir* **25**, 6813–6824 (2009).

[58] Wolthers, M., Charlet, L. & Van Cappellen, P. The surface chemistry of divalent metal carbonate minerals; a critical aassessment of surface charge and potential data using the charge distribution multi-site ion complexation model. *American Journal of Science* **308**, 905–941 (2008).

[59] Somasundaran, P. & Agar, G. The zero point of charge of calcite. *Journal of Colloid and Interface Science* **24**, 433–440 (1967).

[60] van Cappellen, P., Charlet, L., Stumm, W. & Wersin, P. A surface complexation model of the carbonate mineral – aqueous solution interface. *Geochimica et Cosmochimica Acta* **57**, 3505–3518 (1993).

[61] Charlet, L., Wersin, P. & Stumm, W. Surface charge of $MnCO_3$ and $FeCO_3$. *Geochimica et Cosmochimica Acta* **54**, 2329 – 2336 (1990).

Bibliography

[62] Pokrovsky, O. S. & Schott, J. Surface chemistry and dissolution kinetics of divalent metal carbonates. *Environmental Science & Technology* **36**, 426–432 (2002).

[63] Villegas-Jimenez, A., Mucci, A., Pokrovsky, O. S. & Schott, J. Defining reactive sites on hydrated mineral surfaces: Rhombohedral carbonate minerals. *Geochimica et Cosmochimica Acta* **73**, 4326–4345 (2009).

[64] Stipp, S. Toward a conceptual model of the calcite surface: Hydration hydrolysis and surface potential. *Geochimica et Cosmochimica Acta* **63**, 3121–3131 (1999).

[65] Foxall, T., Peterson, G. C., Rendall, H. M. & Smith, A. L. Charge determination at calcium salt / aqueous solution interface. *Journal of the Chemical Society, Faraday Transactions, I.* **75**, 1034–1039 (1979).

[66] Hiemstra, T. & Van Riemsdijk, W. H. A surface structural approach to ion adsorption: The charge distribution (CD) model. *Journal of Colloid and Interface Science* **179**, 488–508 (1996).

[67] Tai, C. Y. & Hsu, H.-P. Crystal growth kinetics of calcite and its comparison with readily soluble salts. *Powder Technology* **121**, 60–67 (2001).

[68] Tai, C. Y., Chien, W. C. & Chen, C. Y. Crystal growth kinetics of calcite in a dense fluidized-bed crystallizer. *AIChE Journal* **45**, 1605–1614 (1999).

[69] Nielsen, A. E. Electrolyte crystal growth mechanisms. *Journal of Crystalgrowth* **67**, 289–2310 (1984).

[70] Rouff, A. A., Elzinga, E. J. & Reeder, R. J. The effect of aging and pH on Pb(II) sorption processes at the calcite - water interface. *Environmental Science & Technology* **40**, 1792–1798 (2006).

[71] Davis, J. A., Fuller, C. C. & Cook, A. D. A model for trace metal sorption processes at the calcite surface: Adsorption of Cd^{2+} and subsequent solid solution formation. *Geochimica et Cosmochimica Acta* **51**, 1477 – 1490 (1987).

[72] Stipp, S. L., Hochella, M. F., Parks, G. A. & Leckie, J. O. Cd^{2+} uptake by calcite, solid-state difusion, and the formation of sold-solution - interface processes observed

Bibliography

with near-surface sensitive techniques (XPS, LEED, and AES). *Geochimica et Cosmochimica Acta* **56**, 1941–1954 (1992).

[73] Stipp, S. L. S., Konnerup-Madsen, J., Franzreb, K., Kulik, A. & Mathieu, H. J. Spontaneous movement of ions through calcite at standard temperature and pressure. *Nature* **396**, 356–359 (1998).

[74] Perez-Garrido, C., Fernandez-Diaz, L., Pina, C. M. & Prieto, M. In situ AFM observations of the interaction between calcite (104) surfaces and Cd-bearing aqueous solutions. *Surface Science* **601**, 5499–5509 (2007).

[75] Astilleros, J. M., Pina, C. M., Fernandez-Diaz, L., Prieto, M. & Putnis, A. Nanoscale phenomena during the growth of solid solutions on calcite (104) surfaces. *Chemical Geology* **225**, 322–335 (2006).

[76] Cherniak, D. J. An experimental study of strontium and lead diffusion in calcite, and implications for carbonate diagenesis and metamorphism. *Geochimica et Cosmochimica Acta* **61**, 4173–4179 (1997).

[77] Elzinga, E. J., Rouff, A. A. & Reeder, R. J. The long-term fate of Cu^{2+}, Zn^{2+}, and Pb^{2+} adsorption complexes at the calcite surface: An x-ray absorption spectroscopy study. *Geochimica et Cosmochimica Acta* **70**, 2715–2725 (2006).

[78] Carroll, S. A., Bruno, J., Petit, J. C. & Dran, J. C. Interactions of U(VI), Nd, and Th(IV) at the calcite-solution interface. *Radiochimica Acta* **58-9**, 245–252 (1992). Part 2.

[79] Piriou, B., Fedoroff, M., Jeanjean, J. & Bercis, L. Characterization of the sorption of europium(III) on calcite by site-selective and time-resolved luminescence spectroscopy. *Journal of Colloid and Interface Science* **194**, 440–447 (1997).

[80] Geipel, G., Reich, T., Brendler, V., Bernhard, G. & Nitsche, H. Laser and x-ray spectroscopic studies of uranium-calcite interface phenomena. *Journal of Nuclear Materials* **248**, 408–411 (1997).

[81] Denecke, M. A. *et al.* Polarization dependent grazing incident (GI) XAFS measurements of uranyl cation sorption onto mineral surfaces. *Physica Scripta* **T115**, 877–881 (2005).

Bibliography

[82] Schmidt, M., Stumpf, T., Fernandes, M. M., Walther, C. & Fanghänel, T. Charge compensation in solid solutions. *Angewandte Chemie-International Edition* **47**, 5846–5850 (2008).

[83] Tesoriero, A. J. & Pankow, J. F. Solid solution partitioning of Sr^{2+}, Ba^{2+}, and Cd^{2+} to calcite. *Geochimica et Cosmochimica Acta* **60**, 1053–1063 (1996).

[84] Glynn, P. D. Solid-solution solubilities and thermodynamics: Sulfates, carbonates, and halides. In Alpers, C. N., Jambor, J. L. & Nordstrom, D. K. (eds.) *Sulfate Minerals: Crystallography, Geochemistry, and Environmental Significance*, vol. 40 of *Reviews in Mineralogy and Geochemistry*, 480–511 (Mineralogical Society of America, 2000).

[85] Shtukenberg, A. G., Punin, Y. O. & Azimov, P. Crystallization kinetics in binary solid solution-aqueous solution systems. *American Journal of Science* **306**, 553–574 (2006).

[86] Guggenheim, E. Theoretical basis of raoult's law. *Transactions of the Faraday Society* **33**, 151–159 (1937).

[87] Lippmann, F. Phase diagrams depicting the aqueous solubility of mineral systems. *Neues Jahrbuch für Mineralogie* **139**, 1–25 (1980).

[88] Pokrovsky, O. S., Mielczarski, J. A., Barres, O. & Scott, J. Surface speciation models of calcite and dolomite / aqueous solution interfaces and their spectroscopic evaluation. *Langmuir* **16**, 2677–2688 (2000).

[89] Newville, M. Fundamentals of xafs. http://cars.uchicago.edu/xafs/ (2004).

[90] Robinson, I. K. Crystal truncation rods and surface roughness. *Physical Review B* **33**, 3830–3836 (1986).

[91] Trainor, T. P., Eng, P. J. & Robinson, I. K. Calculation of crystal truncation rod structure factors for arbitrary rational surface terminations. *Journal of Applied Crystallography* **35**, 696–701 (2002).

[92] You, H. Angle calculations for a '4S+2D' six-circle diffractometer. *Journal of Applied Crystallography* **32**, 614–623 (1999).

Bibliography

[93] GSECARS. http://cars9.uchicago.edu/ifeffit/tdl (2009).

[94] Schlepütz, C. M. *et al.* Improved data acquisition in grazing-incidence X-ray scattering experiments using a pixel detector. *Acta Crystallographica Section A* **61**, 418–425 (2005).

[95] Vlieg, E. ROD: a program for surface x-ray crystallography. *Journal of Applied Crystallography* **33**, 401–405 (2000).

[96] Brown, I. D. & Altermatt, D. Bond-valence parameters obtained from a systematic analysis of the inorganic crystal-structure database. *Acta Crystallographica Section B: Structural Science* **41**, 244 (1985).

[97] Lützenkirchen, J. Surface complexation models of adsorption: A critical survey in the context of experimental data. In *Surface Complexation Modeling*, chap. 11 (Elsevier, Amsterdam, 2006), 1 edn.

[98] Poeter, E. P. & Hill, M. C. Documentation of UCODE, a computer code for universal inverse modeling. Tech. Rep., US Geological Survey (1998).

[99] Westall, J. A computer program for determination of chemical equilibrium constants from experimental data, version 2.0. Tech. Rep., Department of Chemistry, Oregon State University (1982).

[100] Stipp, S. L. & Hochella, M. F. Structure and bonding environments at the calcite surface as observed with x-ray photoelectron spectroscopy (XPS) and low energy electron diffraction (LEED). *Geochimica et Cosmochimica Acta* **55**, 1723–1736 (1991).

[101] Matz, W. *et al.* ROBL - a CRG beamline for radiochemistry and materials research at the ESRF. *Journal of Synchrotron Radiation* **6**, 1076–1085 (1999).

[102] Webb, S. M. SIXpack: a graphical user interface for xas analysis using IFEFFIT. *Physica Scripta* **T115**, 1011–1014 (2005).

[103] Ravel, B. & Newville, M. ATHENA, ARTEMIS, HEPHAESTUS: data analysis for x-ray absorption spectroscopy using ifeffit. *Journal of Synchrotron Radiation* **12**, 537–541 (2005). Part 4.

Bibliography

[104] Ressler, T. WinXAS: A new software package not only for the analysis of energy-dispersive XAS data. *Journal of Physics IV* **7 (C2)**, 269–270 (1997).

[105] Forbes, T. Z., Wallace, C. & Burns, P. C. Neptunyl compounds: Polyhedron geometries, bond-valence parameters, and structural hierarchy. In *Feldspars 2007 Session held at the Frontiers in Mineral Sciences Symposium*, 1623–1645 (Mineralogical Association of Canada, 2007).

[106] Wang, Y. F. & Xu, H. F. Prediction of trace metal partitioning between minerals and aqueous solutions: A linear free energy correlation approach. *Geochimica et Cosmochimica Acta* **65**, 1529–1543 (2001).

[107] Rothe, J., Denecke, M. A., Dardenne, K. & Fanghänel, T. The INE-beamline for actinide research at ANKA. *Radiochimica Acta* **94**, 691–696 (2006).

[108] Rehr, J., de Leon, J. M., Zabinsky, S. & Albers, R. Theoretical x-ray absorption fine structure standards. *Journal of the American Chemical Society* **113**, 5135 (1991).

[109] Hennig, C., Tutschku, J., Rossberg, A., Bernhard, G. & Scheinost, A. C. Comparative EXAFS investigation of uranium(VI) and -(IV) aquo chloro complexes in solution using a newly developed spectroelectrochemical cell. *Inorganic Chemistry Communications* **44**, 6655–6661 (2005).

[110] Ulrich, K.-U., Rossberg, A., Foerstendorf, H., Zanker, H. & Scheinost, A. C. Molecular characterization of uranium(VI) sorption complexes on iron(III)-rich acid mine water colloids. *Geochimica et Cosmochimica Acta* **70**, 5469–5487 (2006).

[111] Guoxin, T., Jide, X. & Linfeng, R. Optical absorption and structure of a highly symmetrical neptunium(V) diamide complex. *Angewandte Chemie* **117**, 6356–6359 (2005).

[112] Clark, D. L. *et al.* A multi-method approach to actinide speciation applied to pentavalent neptunium carbonate complexation. *New Journal of Chemistry* **20**, 211–220 (1996).

[113] Clark, D. L. *et al.* EXAFS studies of pentavalent neptunium carbonato complexes. structural elucidation of principal constituents of neptunium in groundwater environments. *Journal of the American Chemical Society* **118**, 2089–2090 (1996).

Bibliography

[114] Gibbs-Davis, J. M., Kruk, J. J., Konek, C. T., Scheidt, K. A. & Geiger, F. M. Jammed acid base reactions at interfaces. *Journal of the American Chemical Society* **130**, 15444–15447 (2008).

[115] Eriksson, R., Juha, M. & Rosenholm, J. B. The calcite/water interface I. surface charge in indifferent electrolyte media and the influence of low–molecular–weight polyelectrolyte. *Journal of Colloid and Interface Science* **313**, 184–193 (2007).

[116] Hiemstra, T., P., V. & Van Riemsdijk, W. H. Intrinsic proton affinity of reactive surface groups of metal (hydr)oxides: the bond valence principle. *Journal of Colloid and Interface Science* **184**, 680–692 (1996).

[117] Lützenkirchen, J., Preocanin, T. & Kallay, N. A macroscopic water structure based model for describing charging phenomena at inert hydrophobic surfaces in aqueous electrolyte solutions. *Physical Chemistry Chemical Physics* **10**, 4946–4955 (2008).

[118] Morel, M. M. & Hering, J. *Principles and applications of aquatic chemistry* (John Wiley & Sons, New York, 1993), 1 edn.

[119] Heberling, F., Brendebach, B. & Bosbach, D. Neptunium(V) adsorption to calcite. *Journal of Contaminant Hydrology* **102**, 246–252 (2008).

[120] Bernhard, G. *et al.* Uranyl(VI) carbonate complex formation: Validation of the $Ca_2UO_2(CO_3)_{3(aq)}$ species. *Radiochimica Acta* **89**, 511–518 (2001).

[121] Lemire, R. *et al. Chemical Thermodynamics of Neptunium and Plutonium*, vol. 4 of *Chemical Thermodynamics* (Elsevier, Amsterdam, 2001).

[122] Curti, E. Coprecipitation of radionuclides with calcite: estimation of partition coefficients based on a review of laboratory investigations and geochemical data. *Applied Geochemistry* **14**, 433–445 (1999).

[123] Reddy, M. M. & Nancollas, G. H. Calcite crystal growth inhibition by phosphonates. *Desalination* **12**, 61–73 (1973).

[124] Burns, P. C., Ewing, P. C. & Miller, M. L. Incorporation mechanisms of actinide elements into the structures of U^{6+} phases formed during the oxidation of spent nuclear fuel. *Journal of Nuclear Materials* **245**, 1–9 (1997).

Bibliography

[125] White, W. B. *The infra red spectra of minerals* (Mineralogical Society, London, 1974), 1 edn.

[126] Den Auwer, C., Simoni, E., Conradson, S. & Madic, C. Investigating actinyl oxo cations by x-ray absorption spectroscopy. *European Journal of Inorganic Chemistry* 3843–3859 (2003).

[127] Vigfusson, J., Maudoux, J., Raimbault, P., J., R. K. & Smith, R. E. European pilot study on the regulatory review of the safety case for geological disposal of radioactive waste (2007).

Appendix A

ROD files

ROD bulk file

```
CaCO3(104) surface cell with 2 molecular layers
8.0942   4.988   6.0701   90   90   90
```

A. ROD files

O	0.967	0.872	−0.372	2
O	0.467	0.128	−0.372	2
Ca	0.620	0.500	−0.500	1
O	0.870	0.256	−0.500	2
C	0.870	0.000	−0.500	3
C	0.370	1.000	−0.500	3
O	0.370	0.744	−0.500	2
Ca	0.120	0.500	−0.500	1
O	0.773	0.872	−0.628	2
O	0.273	0.128	−0.628	2
O	0.087	0.628	−0.872	2
O	0.587	0.372	−0.872	2
Ca	0.740	0.000	−1.000	1
O	0.490	0.756	−1.000	2
C	0.490	0.500	−1.000	3
C	0.990	0.500	−1.000	3
O	0.990	0.244	−1.000	2
Ca	0.240	0.000	−1.000	1
O	0.393	0.372	−1.128	2
O	0.893	0.628	−1.128	2

A. ROD files

ROD fit file

CaCO3(104) fit file, 3 molecular layers, 4 CO3 groups, delta1 = -0.2405
8.0942 4.988 6.0701 90 90 90
#S   bulk   scell

A. ROD files

O	0.293	1	4	0	0	0.872	-1	5	0	0	1.800	6	11	6 0
O	0.793	1	4	0	0	0.128	1	5	0	0	1.800	6	11	6 0
O	0.607	1	1	0	0	0.628	1	2	0	0	1.800	3	10	5 0
O	1.107	1	1	0	0	0.372	-1	2	0	0	1.800	3	10	5 0
O	1.107	1	22	0	0	0.372	-1	23	0	0	1.500	24	9	4 0
O	0.607	1	22	0	0	0.628	1	23	0	0	1.500	24	9	4 0
O	0.293	1	19	0	0	0.872	-1	20	0	0	1.372	21	8	3 0
O	0.793	1	19	0	0	0.128	1	20	0	0	1.372	21	8	3 0
O	0.607	1	13	0	0	0.628	1	14	0	0	1.128	15	6	1 3
O	1.107	1	13	0	0	0.372	-1	14	0	0	1.128	15	6	1 4
Ca	0.260	1	16	0	0	0.000	-1	17	0	0	1.000	18	7	2 0
O	1.010	1	13	0	0	0.756	-1	14	0	0	1.000	15	6	1 4
C	1.010	1	13	0	0	0.500	-1	14	0	0	1.000	15	6	1 4
C	0.510	1	13	0	0	0.500	1	14	0	0	1.000	15	6	1 3
O	0.510	1	13	0	0	0.244	1	14	0	0	1.000	15	6	1 3
Ca	0.760	1	16	0	0	0.000	1	17	0	0	1.000	18	7	2 0
O	0.913	1	13	0	0	0.372	-1	14	0	0	0.872	15	6	1 4
O	0.413	1	13	0	0	0.628	1	14	0	0	0.872	15	6	1 3
O	0.727	1	7	0	0	-.128	-1	8	0	0	0.628	9	4	0 2
O	0.227	1	7	0	0	1.128	1	8	0	0	0.628	9	4	0 1
Ca	0.380	1	10	0	0	0.500	1	11	0	0	0.500	12	5	0 0
O	0.630	1	7	0	0	0.256	-1	8	0	0	0.500	9	4	0 2
C	0.630	1	7	0	0	0.000	-1	8	0	0	0.500	9	4	0 2
C	0.130	1	7	0	0	1.000	1	8	0	0	0.500	9	4	0 1
O	0.130	1	7	0	0	0.744	1	8	0	0	0.500	9	4	0 1
Ca	0.880	1	10	0	0	0.500	-1	11	0	0	0.500	12	5	0 0
O	0.533	1	7	0	0	-.128	-1	8	0	0	0.372	9	4	0 2
O	0.033	1	7	0	0	1.128	1	8	0	0	0.372	9	4	0 1
O	0.847	1	0	0	0	0.628	1	0	0	0	0.128	0	2	0 0
O	0.347	1	0	0	0	0.372	-1	0	0	0	0.128	0	2	0 0

A. ROD files

```
Ca   0.500  1  0  0 0   0.000  -1  0  0 0   0.000   0  1  0 0
O    0.250  1  0  0 0   0.756  -1  0  0 0   0.000   0  2  0 0
C    0.250  1  0  0 0   0.500  -1  0  0 0   0.000   0  3  0 0
C    0.750  1  0  0 0   0.500   1  0  0 0   0.000   0  3  0 0
O    0.750  1  0  0 0   0.244   1  0  0 0   0.000   0  2  0 0
Ca   0.000  1  0  0 0   0.000   1  0  0 0   0.000   0  1  0 0
O    0.153  1  0  0 0   0.372  -1  0  0 0  -.128    0  2  0 0
O    0.653  1  0  0 0   0.628   1  0  0 0  -.128    0  2  0 0
#G   1  group
0.130    1.00000   0.5000
0  1  1  90  1  2  0  1  3
#G   2  group
0.630    0.00001   0.5000
0 -1  1 -90 -1  2  0 -1  3
#G   3  group
0.510   0.50000   1.0000
0  1  4  90  1  5  0  1  6
#G   4  group
1.010   0.50000   1.0000
0 -1  4 -90 -1  5  0 -1  6
```

Die VDM Verlagsservicegesellschaft sucht für wissenschaftliche Verlage abgeschlossene und herausragende

Dissertationen, Habilitationen, Diplomarbeiten, Master Theses, Magisterarbeiten usw.

für die kostenlose Publikation als Fachbuch.

Sie verfügen über eine Arbeit, die hohen inhaltlichen und formalen Ansprüchen genügt, und haben Interesse an einer honorarvergüteten Publikation?

Dann senden Sie bitte erste Informationen über sich und Ihre Arbeit per Email an *info@vdm-vsg.de*.

Sie erhalten kurzfristig unser Feedback!

VDM Verlagsservicegesellschaft mbH
Dudweiler Landstr. 99   Telefon +49 681 3720 174
D - 66123 Saarbrücken   Fax   +49 681 3720 1749
www.vdm-vsg.de

Die VDM Verlagsservicegesellschaft mbH vertritt

Printed by Books on Demand GmbH, Norderstedt / Germany